T0331684

Exercises and Problems in
Linear Algebra

Exercises and Problems in
Linear Algebra

John M. Erdman
Portland State University, USA

 **World Scientific**

NEW JERSEY · LONDON · SINGAPORE · BEIJING · SHANGHAI · HONG KONG · TAIPEI · CHENNAI · TOKYO

Published by

World Scientific Publishing Co. Pte. Ltd.

5 Toh Tuck Link, Singapore 596224

USA office: 27 Warren Street, Suite 401-402, Hackensack, NJ 07601

UK office: 57 Shelton Street, Covent Garden, London WC2H 9HE

Library of Congress Cataloging-in-Publication Data
Names: Erdman, John M., 1935–   author.
Title: Exercises and problems in linear algebra / John M. Erdman, Portland State University.
Description: New Jersey : World Scientific, [2021] | Includes bibliographical references and index.
Identifiers: LCCN 2020038857 (print) | LCCN 2020038858 (ebook) |
    ISBN 9789811220401 (hardcover) | ISBN 9789811221071 (paperback) |
    ISBN 9789811220418 (ebook) | ISBN 9789811220425 (ebook other)
Subjects: LCSH: Algebras, Linear--Problems, exercises, etc.
Classification: LCC QA179 .E73 2021 (print) | LCC QA179 (ebook) | DDC 512/.5--dc23
LC record available at https://lccn.loc.gov/2020038857
LC ebook record available at https://lccn.loc.gov/2020038858

British Library Cataloguing-in-Publication Data
A catalogue record for this book is available from the British Library.

For any available supplementary material, please visit
https://www.worldscientific.com/worldscibooks/10.1142/11830#t=suppl

Desk Editor: Soh Jing Wen

Typeset by Stallion Press
Email: enquiries@stallionpress.com

Printed in Singapore

CONTENTS

PREFACE

This collection of exercises and problems may be used in many ways. I used it as a source of discussion topics in an activity-oriented course in linear algebra. The discussions were based principally on student presentations of solutions to the items included in this study guide and on the difficulties they encountered while trying to solve them. There were no assigned texts. Students were free to choose their own sources of information and were encouraged to consult books, papers, and websites whose writing style they found congenial, whose emphasis matched their interests, and whose prices fit their budgets.

The exercises, I hope, will be both interesting and helpful to an average student. Some are fairly routine calculations, while others require serious thought. The easy-to-grade fill-in-the-blank format makes them suitable for instructors to use in quizzes and assigned homework. The answers, given for all odd-numbered exercises, should be useful for self-directed learners.

Most of the problems were designed to illustrate some issues that I regard as needing emphasis. Some of them are somewhat open-ended and are suitable for projects in classes where groups of students work together. For those instructors who favor the lecture method or are forced into it by having classes so large that class discussions or group projects are only remote fantasies, the problems herein may provide excellent topics for presentation. As an added convenience for instructors the LaTeX code for all the exercises and problems will be made available on the *World Scientific* website. This should be helpful since some of these items are complicated and/or lengthy and would be a nuisance to retype for use on exams or homework assignments.

In each chapter there is a short introductory *background* section that lists the topics on which the following exercises and problems are based. Since notation and terminology in linear algebra are not entirely uniform throughout the literature, these background sections are intended to fix notation and provide "official" definitions and statements of important theorems for the material that follows. It seems clear that students in a class communicate better if they are encouraged to use agreed-upon notations and terminology.

One problem that I find with many introductory linear algebra texts is their conflation of material on vector spaces with that on inner product spaces. The subjects differ in both notation and terminology. Terms such as "direct sum" and "projection" mean one thing in one context and something quite different in the other. In the present study guide I try to keep the algebraic business about vector spaces (where words like "length", "angle", "distance", and "perpendicular" make no sense) separate from the more geometric material on inner products. In this study guide vector spaces are dealt with in parts 2–4, while inner product spaces are the subject of parts 5–7.

Among the dozens and dozens of linear algebra texts that have appeared, two that were written before "dumbing down" of textbooks became fashionable are especially notable, in my opinion, for the clarity of their authors' mathematical vision: Paul Halmos's *Finite-Dimensional Vector Spaces* [4] and Hoffman and Kunze's *Linear Algebra* [6]. Some students, especially mathematically inclined ones, love these books, but others find them challenging. For students trying seriously to learn the subject, I recommend examining them. For those more interested in applications, both *Elementary Linear Algebra: Applications Version* [1] by Howard Anton and Chris Rorres and *Linear Algebra and its Applications* [9] by Gilbert Strang are loaded with applications. Students who find the level of many of the current beginning linear algebra texts depressingly pedestrian and the endless routine computations irritating, may enjoy reading some more advanced texts. Two excellent ones are Steven Roman's *Advanced Linear Algebra* [8] and William C. Brown's *A Second Course in Linear Algebra* [2].

Concerning the material in these notes, I make no claims of originality. While I have dreamed up many of the items included here, there are many others which are standard linear algebra exercises that can be traced back, in one form or another, through generations of linear algebra texts, making any serious attempt at proper attribution quite futile. If anyone feels slighted, please contact me.

There will surely be errors. I will be delighted to receive corrections, suggestions, or criticism at

erdmanj@comcast.net

Part 1

MATRICES AND LINEAR EQUATIONS

Chapter 1

·

ARITHMETIC OF MATRICES

1.1 Background

Topics: addition, scalar multiplication, multiplication of matrices, inverse of a nonsingular matrix, Gaussian elimination.

Definition 1.1.1. Two square matrices A and B of the same size are said to COMMUTE if $AB = BA$.

Definition 1.1.2. If A and B are square matrices of the same size, then the COMMUTATOR (or LIE BRACKET) of A and B, denoted by $[A, B]$, is defined by

$$[A, B] = AB - BA.$$

Notation 1.1.3. If A is an $m \times n$ matrix (that is, a matrix with m rows and n columns), then the element in the ith row and the jth column is denoted by a_{ij}. The matrix A itself may be denoted by $\left[a_{ij}\right]_{i=1\,j=1}^{m\quad n}$ or, more simply, by $[a_{ij}]$. In light of this notation it is reasonable to refer to the index i in the expression a_{ij} as the ROW INDEX and to call j the COLUMN INDEX. When we speak of the "value of a matrix A at (i, j)," we mean the entry in the ith row and j^{th} column of A. Thus, for example,

$$A = \begin{bmatrix} 1 & 4 \\ 3 & -2 \\ 7 & 0 \\ 5 & -1 \end{bmatrix}$$

is a 4×2 matrix and $a_{31} = 7$.

3

Notation 1.1.4. The $n \times n$ identity matrix is denoted by I_n or, more frequently, just by I.

Definition 1.1.5. A matrix $A = [a_{ij}]$ is UPPER TRIANGULAR if $a_{ij} = 0$ whenever $i > j$.

Definition 1.1.6. The TRACE of a square matrix A, denoted by $\operatorname{tr} A$, is the sum of the diagonal entries of the matrix. That is, if $A = [a_{ij}]$ is an $n \times n$ matrix, then

$$\operatorname{tr} A := \sum_{j=1}^{n} a_{jj}.$$

Definition 1.1.7. The TRANSPOSE of an $m \times n$ matrix $A = [a_{ij}]$ is the matrix $A^t = [a_{ji}]$ obtained by interchanging the rows and columns of A. The matrix A is SYMMETRIC if $A^t = A$.

Proposition 1.1.8. *If A is an $m \times n$ matrix and B is an $n \times p$ matrix, then $(AB)^t = B^t A^t$.*

1.2   Exercises

(1) Let $A = \begin{bmatrix} 1 & 0 & -1 & 2 \\ 0 & 3 & 1 & -1 \\ 2 & 4 & 0 & 3 \\ -3 & 1 & -1 & 2 \end{bmatrix}$, $B = \begin{bmatrix} 1 & 2 \\ 3 & -1 \\ 0 & -2 \\ 4 & 1 \end{bmatrix}$, and $C = \begin{bmatrix} 3 & -2 & 0 & 5 \\ 1 & 0 & -3 & 4 \end{bmatrix}$.

   (a) Does the matrix $D = ABC$ exist? _____ If so, then $d_{34} = $ _____.

   (b) Does the matrix $E = BAC$ exist? _____ If so, then $e_{22} = $ _____.

   (c) Does the matrix $F = BCA$ exist? _____ If so, then $f_{43} = $ _____.

   (d) Does the matrix $G = ACB$ exist? _____ If so, then $g_{31} = $ _____.

   (e) Does the matrix $H = CAB$ exist? _____ If so, then $h_{21} = $ _____.

   (f) Does the matrix $J = CBA$ exist? _____ If so, then $j_{13} = $ _____.

(2) Let $A = \begin{bmatrix} \frac{1}{2} & \frac{1}{2} \\ \frac{1}{2} & \frac{1}{2} \end{bmatrix}$, $B = \begin{bmatrix} 1 & 0 \\ 0 & -1 \end{bmatrix}$, and $C = AB$. Evaluate the following.

(a) $A^{37} = \begin{bmatrix} & \\ & \end{bmatrix}$

(b) $B^{63} = \begin{bmatrix} & \\ & \end{bmatrix}$

(c) $B^{138} = \begin{bmatrix} & \\ & \end{bmatrix}$

(d) $C^{42} = \begin{bmatrix} & \\ & \end{bmatrix}$

Note: If M is a matrix M^p is the product of p copies of M.

(3) Let $A = \begin{bmatrix} 1 & 1/3 \\ c & d \end{bmatrix}$. Find numbers c and d such that $A^2 = -I$.

Answer: $c =$ _____ and $d =$ _____.

(4) Let A and B be symmetric $n \times n$-matrices. Then $[A, B] = [B, X]$, where $X =$ _____.

(5) Let A, B, and C be $n \times n$ matrices. Then $[A, B]C + B[A, C] = [X, Y]$, where $X =$ _____ and $Y =$ _____.

(6) Let $A = \begin{bmatrix} 1 & 1/3 \\ c & d \end{bmatrix}$. Find numbers c and d such that $A^2 = 0$. Answer: $c =$ _____ and $d =$ _____.

(7) Consider the matrix $\begin{bmatrix} 1 & 3 & 2 \\ a & 6 & 2 \\ 0 & 9 & 5 \end{bmatrix}$ where a is a real number.

(a) For what value of a will a row interchange be required during Gaussian elimination? Answer: $a =$ _____.

(b) For what value of a is the matrix singular? Answer: $a =$ _____.

(8) Let $A = \begin{bmatrix} 1 & 0 & -1 & 2 \\ 0 & 3 & 1 & -1 \\ 2 & 4 & 0 & 3 \\ -3 & 1 & -1 & 2 \end{bmatrix}$, $B = \begin{bmatrix} 1 & 2 \\ 3 & -1 \\ 0 & -2 \\ 4 & 1 \end{bmatrix}$, $C = \begin{bmatrix} 3 & -2 & 0 & 5 \\ 1 & 0 & -3 & 4 \end{bmatrix}$,

and $M = 3A^3 - 5(BC)^2$. Then $m_{14} =$ _____ and $m_{41} =$ _____.

(9) If A is an $n \times n$ matrix and it satisfies the equation $A^3 - 4A^2 + 3A - 5I_n = 0$, then A is nonsingular and its inverse is

_____.

(10) Let A, B, and C be $n \times n$ matrices. Then $[[A, B], C] + [[B, C], A] + [[C, A], B] = X$, where $X = \begin{bmatrix} \\ \\ \end{bmatrix}$.

(11) Let A, B, and C be $n \times n$ matrices. Then $[A, C] + [B, C] = [X, Y]$, where $X = $ _____ and $Y = $ _____.

(12) Find the inverse of $\begin{bmatrix} 1 & 0 & 0 & 0 \\ \frac{1}{4} & 1 & 0 & 0 \\ \frac{1}{3} & \frac{1}{3} & 1 & 0 \\ \frac{1}{2} & \frac{1}{2} & \frac{1}{2} & 1 \end{bmatrix}$. Answer: $\begin{bmatrix} \\ \\ \\ \end{bmatrix}$.

(13) Suppose that A and B are symmetric $n \times n$ matrices. In this exercise we prove that AB is symmetric if and only if A commutes with B. Below are portions of the proof. Fill in the missing steps and the missing reasons. Choose reasons from the following list.

> (H1) Hypothesis that A and B are symmetric.
>
> (H2) Hypothesis that AB is symmetric.
>
> (H3) Hypothesis that A commutes with B.
>
> (D1) Definition of *commutes*.
>
> (D2) Definition of *symmetric*.
>
> (T) Proposition 1.1.8.

Proof. Suppose that AB is symmetric. Then

$$AB = \underline{\hspace{2cm}} \qquad \text{(reason: (H2) and } \underline{\hspace{1cm}})$$
$$= B^t A^t \qquad \text{(reason: } \underline{\hspace{1cm}})$$
$$= \underline{\hspace{2cm}} \qquad \text{(reason: (D2) and } \underline{\hspace{1cm}})$$

So A commutes with B (reason: _____).

Conversely, suppose that A commutes with B. Then

$$(AB)^t = \underline{\hspace{2cm}} \qquad \text{(reason: (T))}$$
$$= BA \qquad \text{(reason: } \underline{\hspace{1.5cm}} \text{ and } \underline{\hspace{1.5cm}} \text{)}$$
$$= \underline{\hspace{2cm}} \qquad \text{(reason: } \underline{\hspace{1.5cm}} \text{ and } \underline{\hspace{1.5cm}} \text{)}$$

Thus AB is symmetric (reason: $\underline{\hspace{2cm}}$). $\square$

1.3 Problems

(1) Let A be a square matrix. Prove that if A^2 is invertible, then so is A.
Hint. Our assumption is that there exists a matrix B such that

$$A^2 B = BA^2 = I.$$

We want to show that there exists a matrix C such that

$$AC = CA = I.$$

Now to start with, you ought to find it fairly easy to show that there are matrices L and R such that

$$LA = AR = I. \tag{$*$}$$

A matrix L is a LEFT INVERSE of the matrix A if $LA = I$; and R is a RIGHT INVERSE of A if $AR = I$. Thus the problem boils down to determining whether A can have a left inverse and a right inverse that are *different*. (Clearly, if it turns out that they must be the same, then the C we are seeking is their common value.) So try to prove that if $(*)$ holds, then $L = R$.

(2) Anton speaks French and German; Geraldine speaks English, French and Italian; James speaks English, Italian, and Spanish; Lauren speaks all the languages the others speak except French; and no one speaks any other language. Make a matrix $A = [a_{ij}]$ with rows representing the four people mentioned and columns representing the languages they speak. Put $a_{ij} = 1$ if person i speaks language j and $a_{ij} = 0$ otherwise. Explain the significance of the matrices AA^t and $A^t A$.

(3) Portland Fast Foods (PFF), which produces 138 food products all made from 87 basic ingredients, wants to set up a simple data structure from

which they can quickly extract answers to the following questions:

(a) How many ingredients does a given product contain?

(b) A given pair of ingredients are used together in how many products?

(c) How many ingredients do two given products have in common?

(d) In how many products is a given ingredient used?

In particular, PFF wants to set up a single table in such a way that:

(i) the answer to any of the above questions can be extracted easily and quickly (matrix arithmetic permitted, of course); and

(ii) if one of the 87 ingredients is added to or deleted from a product, only a single entry in the table needs to be changed.

Is this possible? Explain.

(4) Prove Proposition 1.1.8.

(5) Let A and B be 2×2 matrices.

(a) Prove that if the trace of A is 0, then A^2 is a scalar multiple of the identity matrix.

(b) Prove that the square of the commutator of A and B commutes with every 2×2 matrix C. *Hint.* What can you say about the trace of $[A, B]$?

(c) Prove that the commutator of A and B can never be a nonzero multiple of the identity matrix.

(6) The matrices that represent rotations of the xy-plane are

$$A(\theta) = \begin{bmatrix} \cos \theta & -\sin \theta \\ \sin \theta & \cos \theta \end{bmatrix}.$$

(a) Let $\mathbf{x}$ be the vector $(-1, 1)$, $\theta = 3\pi/4$, and $\mathbf{y}$ be $A(\theta)$ acting on $\mathbf{x}$ (that is, $\mathbf{y} = A(\theta)\mathbf{x}^t$). Make a sketch showing $\mathbf{x}$, $\mathbf{y}$, and θ.

(b) Verify that $A(\theta_1)A(\theta_2) = A(\theta_1 + \theta_2)$. Discuss what this means geometrically.

(c) What is the product of $A(\theta)$ times $A(-\theta)$? Discuss what this means geometrically.

(d) Two sheets of graph paper are attached at the origin and rotated in such a way that the point $(1, 0)$ on the upper sheet lies directly over the point $(-5/13, 12/13)$ on the lower sheet. What point on the lower sheet lies directly below $(6, 4)$ on the upper one?

(7) Let

$$A = \begin{bmatrix} 0 & a & a^2 & a^3 & a^4 \\ 0 & 0 & a & a^2 & a^3 \\ 0 & 0 & 0 & a & a^2 \\ 0 & 0 & 0 & 0 & a \\ 0 & 0 & 0 & 0 & 0 \end{bmatrix}.$$

The goal of this problem is to develop a "calculus" for the matrix A. To start, recall (or look up) the power series expansion for $\dfrac{1}{1-x}$. Now see if this formula works for the $n \times m$ matrix A by first computing $(I-A)^{-1}$ directly and then computing the power series expansion substituting A for x. (Explain why there are no convergence difficulties for the series when we use this particular matrix A.) Next try to define $\ln(I+A)$ and e^A by means of appropriate series. Do you get what you expect when you compute $e^{\ln(I+A)}$? Do formulas like $e^A e^A = e^{2A}$ hold? What about other familiar properties of the exponential and logarithmic functions?

Try some trigonometry with A. Use series to define sin, cos, tan, arctan, and so on. Do things like $\tan(\arctan(A))$ produce the expected results? Check some of the more obvious trigonometric identities. (What do you get for $\sin^2 A + \cos^2 A - I$? Is $\cos(2A)$ the same as $\cos^2 A - \sin^2 A$?)

A relationship between the exponential and trigonometric functions is given by the famous formula $e^{ix} = \cos x + i \sin x$. Does this hold for A?

Do you think there are other matrices for which the same results might hold? Which ones?

(8) (a) Give an example of two symmetric matrices whose product is not symmetric.

   Hint. Matrices containing only 0's and 1's will suffice.

   (b) Now suppose that A and B are symmetric $n \times n$ matrices. Prove that AB is symmetric if and only if A commutes with B.

   Hint. To prove that a statement P holds "if and only if" a statement Q holds you must first show that P implies Q and then show that Q implies P. In the current problem, there are 4 conditions to be considered:

   (i) $A^t = A$ (A is symmetric),
   (ii) $B^t = B$ (B is symmetric),
   (iii) $(AB)^t = AB$ (AB is symmetric), and
   (iv) $AB = BA$ (A commutes with B).

Recall also the fact given in

(v) Proposition 1.1.8.

The first task is to derive (iv) from (i), (ii), (iii), and (v). Then try to derive (iii) from (i), (ii), (iv), and (v).

1.4   Answers to Odd-Numbered Exercises

(1) (a) yes, 142

  (b) no, −

  (c) yes, −45

  (d) no, −

  (e) yes, −37

  (f) no, −

(3) −6, −1

(5) A, BC

(7) (a) 2

  (b) −4

(9) $\frac{1}{5}(A^2 - 4A + 3I_n)$

(11) $A + B$, C

(13) $(AB)^t$, D2, T, BA, H1, D1, $B^t A^t$, H1, D2, AB, H3, D1, D2 (Note: the order of H1 and D2 and the order of H3 and D1 may be reversed.)

Chapter 2

ELEMENTARY MATRICES; DETERMINANTS

2.1 Background

Topics: elementary (reduction) matrices, determinants, Gauss-Jordan reduction.

The following definition says that we often regard the effect of multiplying a matrix M on the left by another matrix A as the *action of A on M*.

Definition 2.1.1. We say that the matrix A ACTS ON the matrix M to produce the matrix N if $N = AM$. For example the matrix $\begin{bmatrix} 0 & 1 \\ 1 & 0 \end{bmatrix}$ acts on any 2×2 matrix by interchanging (swapping) its rows because $\begin{bmatrix} 0 & 1 \\ 1 & 0 \end{bmatrix} \begin{bmatrix} a & b \\ c & d \end{bmatrix} = \begin{bmatrix} c & d \\ a & b \end{bmatrix}$.

Definition 2.1.2. We will say that an operation (sometimes called *scaling*) that multiplies a row of a matrix (or an equation) by a nonzero constant is a ROW OPERATION OF TYPE I. An operation (sometimes called *swapping*) that interchanges two rows of a matrix (or two equations) is a ROW OPERATION OF TYPE II. And an operation (sometimes called *pivoting*) that adds a multiple of one row of a matrix to another row (or adds a multiple of one equation to another) is a ROW OPERATION OF TYPE III.

Notation 2.1.3. We adopt the following notation for elementary matrices that implement type I row operations. Let A be a matrix having n rows. For any real number $r \neq 0$ denote by $M_j(r)$ the $n \times n$ matrix that acts on A by multiplying its jth row by r. (See exercise 1.)

Notation 2.1.4. We use the following notation for elementary matrices that implement type II row operations. (See Definition 2.1.2.) Let A be a matrix having n rows. Denote by P_{ij} the $n \times n$ matrix that acts on A by interchanging its ith and j^{th} rows. (See exercise 2.)

Notation 2.1.5. And we use the following notation for elementary matrices that implement type III row operations. (See Definition 2.1.2.) Let A be a matrix having n rows. For any real number r denote by $E_{ij}(r)$ the $n \times n$ matrix that acts on A by adding r times the jth row of A to the ith row. (See exercise 3.)

Definition 2.1.6. If a matrix B can be produced from a matrix A by a sequence of elementary row operations, then A and B are ROW EQUIVALENT.

Some Facts about Determinants

Proposition 2.1.7. *Let $n \in \mathbb{N}$ and $\mathbf{M}_{n \times n}$ be the collection of all $n \times n$ matrices. There is exactly one function*

$$\det \colon \mathbf{M}_{n \times n} \to \mathbb{R} \colon A \mapsto \det A$$

which satisfies

(a) *$\det I_n = 1$.*
(b) *If $A \in \mathbf{M}_{n \times n}$ and A' is the matrix obtained by interchanging two rows of A, then $\det A' = -\det A$.*
(c) *If $A \in \mathbf{M}_{n \times n}$, $c \in \mathbb{R}$, and A' is the matrix obtained by multiplying each element in one row of A by the number c, then $\det A' = c \det A$.*
(d) *If $A \in \mathbf{M}_{n \times n}$, $c \in \mathbb{R}$, and A' is the matrix obtained from A by multiplying one row of A by c and adding it to another row of A (that is, choose i and j between 1 and n with $i \neq j$ and replace a_{jk} by $a_{jk} + c a_{ik}$ for $1 \leq k \leq n$), then $\det A' = \det A$.*

Definition 2.1.8. The unique function $\det \colon \mathbf{M}_{n \times n} \to \mathbb{R}$ described above is the $n \times n$ DETERMINANT FUNCTION.

Proposition 2.1.9. *If $A = [a]$ for $a \in \mathbb{R}$ (that is, if $A \in \mathbf{M}_{1 \times 1}$), then* $\det A = a$; *if $A \in \mathbf{M}_{2 \times 2}$, then*

$$\det A = a_{11}a_{22} - a_{12}a_{21}.$$

Proposition 2.1.10. *If $A, B \in \mathbf{M}_{n \times n}$, then $\det(AB) = (\det A)(\det B)$.*

Proposition 2.1.11. *If $A \in \mathbf{M}_{n \times n}$, then $\det A^t = \det A$. (An obvious corollary of this: in conditions* (b), (c), *and* (d) *of Proposition 2.1.7 the word "columns" may be substituted for the word "rows".)*

Definition 2.1.12. Let A be an $n \times n$ matrix. The MINOR of the element a_{jk}, denoted by M_{jk}, is the determinant of the $(n-1) \times (n-1)$ matrix which results from the deletion of the jth row and k^{th} column of A. The COFACTOR of the element a_{jk}, denoted by C_{jk} is defined by

$$C_{jk} := (-1)^{j+k} M_{jk}.$$

Proposition 2.1.13. *If $A \in \mathbf{M}_{n \times n}$ and $1 \leq j \leq n$, then*

$$\det A = \sum_{k=1}^{n} a_{jk} C_{jk}.$$

This is the (LAPLACE) EXPANSION *of the determinant along the jth row.*

In light of 2.1.11, it is clear that expansion along columns works as well as expansion along rows. That is,

$$\det A = \sum_{j=1}^{n} a_{jk} C_{jk}$$

for any k between 1 and n. This is the (LAPLACE) EXPANSION of the determinant along the kth column.

Proposition 2.1.14. *An $n \times n$ matrix A is invertible if and only if $\det A \neq 0$. If A is invertible, then*

$$A^{-1} = (\det A)^{-1} C^t$$

where $C = [C_{jk}]$ is the matrix of cofactors of elements of A.

Definition 2.1.15. A square matrix is SINGULAR if its determinant is zero. Otherwise it is NONSINGULAR.

2.2   Exercises

(1) Let A be a matrix with 4 rows. The matrix $M_3(4)$ that multiplies

the 3rd row of A by 4 is $\begin{bmatrix} & & \\ & & \\ & & \end{bmatrix}$. (See 2.1.3.)

(2) Let A be a matrix with 4 rows. The matrix P_{24} that interchanges

the 2nd and 4th rows of A is $\begin{bmatrix} & & \\ & & \\ & & \end{bmatrix}$. (See 2.1.4.)

(3) Let A be a matrix with 4 rows. The matrix $E_{23}(-2)$ that adds -2 times

the 3rd row of A to the 2nd row is $\begin{bmatrix} & & \\ & & \\ & & \end{bmatrix}$. (See 2.1.5.)

(4) Let A be the 4×4 elementary matrix $E_{43}(-6)$. Then $A^{11} =$

$\begin{bmatrix} & & \\ & & \\ & & \end{bmatrix}$ and $A^{-9} = \begin{bmatrix} & & \\ & & \\ & & \end{bmatrix}$.

(5) Let B be the elementary 4×4 matrix P_{24}. Then $B^{-9} =$

$\begin{bmatrix} & & \\ & & \\ & & \end{bmatrix}$ and $B^{10} = \begin{bmatrix} & & \\ & & \\ & & \end{bmatrix}$.

(6) Let C be the elementary 4×4 matrix $M_3(-2)$. Then $C^4 =$

$\begin{bmatrix} & & \\ & & \\ & & \end{bmatrix}$ and $C^{-3} = \begin{bmatrix} & & \\ & & \\ & & \end{bmatrix}$.

(7) Let $A = \begin{bmatrix} 1 & 2 & 3 \\ 0 & -1 & 1 \\ -2 & 1 & 0 \\ -1 & 2 & -3 \end{bmatrix}$ and $B = P_{23}E_{34}(-2)M_3(-2)E_{42}(1)P_{14}A$.

Then $b_{23} = \underline{\hspace{1cm}}$ and $b_{32} = \underline{\hspace{1cm}}$.

(8) Elementary row operations are performed on a 4×4 matrix A to bring it to upper triangular form. The result is

$$P_{24}E_{43}\left(\frac{7}{2}\right)M_3(5)E_{42}(2)E_{31}(1)E_{21}(3)A = \begin{bmatrix} 1 & 2 & -2 & 0 \\ 0 & -1 & 0 & 1 \\ 0 & 0 & -2 & 2 \\ 0 & 0 & 0 & 10 \end{bmatrix}.$$

Then the determinant of A is _____.

(9) The system of equations:

$$\begin{cases} 2y + 3z = 7 \\ x + y - z = -2 \\ -x + y - 5z = 0 \end{cases}$$

is solved by applying Gauss-Jordan reduction to the augmented coefficient matrix $A = \begin{bmatrix} 0 & 2 & 3 & 7 \\ 1 & 1 & -1 & -2 \\ -1 & 1 & -5 & 0 \end{bmatrix}$. Give the names of the elementary 3×3 matrices $X_1, \ldots, X_8$ that implement the following reduction.

$$A \xrightarrow{X_1} \begin{bmatrix} 1 & 1 & -1 & -2 \\ 0 & 2 & 3 & 7 \\ -1 & 1 & -5 & 0 \end{bmatrix} \xrightarrow{X_2} \begin{bmatrix} 1 & 1 & -1 & -2 \\ 0 & 2 & 3 & 7 \\ 0 & 2 & -6 & -2 \end{bmatrix}$$

$$\xrightarrow{X_3} \begin{bmatrix} 1 & 1 & -1 & -2 \\ 0 & 2 & 3 & 7 \\ 0 & 0 & -9 & -9 \end{bmatrix} \xrightarrow{X_4} \begin{bmatrix} 1 & 1 & -1 & -2 \\ 0 & 2 & 3 & 7 \\ 0 & 0 & 1 & 1 \end{bmatrix}$$

$$\xrightarrow{X_5} \begin{bmatrix} 1 & 1 & -1 & -2 \\ 0 & 2 & 0 & 4 \\ 0 & 0 & 1 & 1 \end{bmatrix} \xrightarrow{X_6} \begin{bmatrix} 1 & 1 & -1 & -2 \\ 0 & 1 & 0 & 2 \\ 0 & 0 & 1 & 1 \end{bmatrix}$$

$$\xrightarrow{X_7} \begin{bmatrix} 1 & 1 & 0 & -1 \\ 0 & 1 & 0 & 2 \\ 0 & 0 & 1 & 1 \end{bmatrix} \xrightarrow{X_8} \begin{bmatrix} 1 & 0 & 0 & -3 \\ 0 & 1 & 0 & 2 \\ 0 & 0 & 1 & 1 \end{bmatrix}.$$

Answer: $X_1 =$ _____, $X_2 =$ _____, $X_3 =$ _____, $X_4 =$ _____, $X_5 =$ _____, $X_6 =$ _____, $X_7 =$ _____, $X_8 =$ _____.

(10) Solve the following equation for x:

$$\det \begin{bmatrix} 3 & -4 & 7 & 0 & 6 & -2 \\ 2 & 0 & 1 & 8 & 0 & 0 \\ 3 & 4 & -8 & 3 & 1 & 2 \\ 27 & 6 & 5 & 0 & 0 & 3 \\ 3 & x & 0 & 2 & 1 & -1 \\ 1 & 0 & -1 & 3 & 4 & 0 \end{bmatrix} = 0. \qquad \text{Answer: } x = \underline{\quad}.$$

(11) Let $A = \begin{bmatrix} 0 & 0 & 1 \\ 0 & 2 & 4 \\ 1 & 2 & 3 \end{bmatrix}$. Find A^{-1} using the technique of augmenting
A by the identity matrix I and performing Gauss-Jordan reduction
on the augmented matrix. The reduction can be accomplished by the
application of five elementary 3×3 matrices. Find elementary matrices
X_1, X_2, and X_3 such that $A^{-1} = X_3 E_{13}(-3) X_2 M_2(1/2) X_1 I$.

(a) The required matrices are $X_1 = P_{1i}$ where $i = \underline{\quad}$, $X_2 = E_{jk}(-2)$ where $j = \underline{\quad}$ and $k = \underline{\quad}$, and $X_3 = E_{12}(r)$ where $r = \underline{\quad}$.

(b) And then $A^{-1} = \begin{bmatrix} & & \\ & & \\ & & \end{bmatrix}$.

(12) $\det \begin{bmatrix} 1 & t & t^2 & t^3 \\ t & 1 & t & t^2 \\ t^2 & t & 1 & t \\ t^3 & t^2 & t & 1 \end{bmatrix} = (1 - a(t))^p$ where $a(t) = \underline{\quad}$ and
$p = \underline{\quad}$.

(13) Evaluate each of the following determinants.

(a) $\det \begin{bmatrix} 6 & 9 & 39 & 49 \\ 5 & 7 & 32 & 37 \\ 3 & 4 & 4 & 4 \\ 1 & 1 & 1 & 1 \end{bmatrix} = \underline{\quad}$.

(b) $\det \begin{bmatrix} 1 & 0 & 1 & 1 \\ 1 & -1 & 2 & 0 \\ 2 & -1 & 3 & 1 \\ 4 & 17 & 0 & -5 \end{bmatrix} = \underline{\quad}$.

(c) $\det \begin{bmatrix} 13 & 3 & -8 & 6 \\ 0 & 0 & -4 & 0 \\ 1 & 0 & 7 & -2 \\ 3 & 0 & 2 & 0 \end{bmatrix} = \underline{\hspace{1cm}}.$

(14) Let M be the matrix $\begin{bmatrix} 5 & 4 & -2 & 3 \\ 5 & 7 & -1 & 8 \\ 5 & 7 & 6 & 10 \\ 5 & 7 & 1 & 9 \end{bmatrix}.$

(a) The determinant of M can be expressed as the constant 5 times the determinant of the single 3×3 matrix $\begin{bmatrix} 3 & 1 & 5 \\ 3 & & \\ 3 & & \end{bmatrix}.$

(b) The determinant of this 3×3 matrix can be expressed as the constant 3 times the determinant of the single 2×2 matrix $\begin{bmatrix} 7 & 2 \\ 2 & \end{bmatrix}.$

(c) The determinant of this 2×2 matrix is ____.

(d) Thus the determinant of M is ____.

(15) Find the determinant of the matrix $\begin{bmatrix} 1 & 2 & 5 & 7 & 10 \\ 1 & 2 & 3 & 6 & 7 \\ 1 & 1 & 3 & 5 & 5 \\ 1 & 1 & 2 & 4 & 5 \\ 1 & 1 & 1 & 1 & 1 \end{bmatrix}.$ Answer: _____.

(16) Find the determinants of the following matrices.

$$A = \begin{bmatrix} -73 & 78 & 24 \\ 92 & 66 & 25 \\ -80 & 37 & 10 \end{bmatrix} \quad \text{and} \quad B = \begin{bmatrix} -73 & 78 & 24 \\ 92 & 66 & 25 \\ -80 & 37 & 10.01 \end{bmatrix}.$$

Hint. Use a calculator (thoughtfully). Answer: $\det A =$ _____ and $\det B =$ _____.

(17) Find the determinant of the following matrix.

$$\begin{bmatrix} 283 & 5 & \pi & 347.86 \times 10^{15^{83}} \\ 3136 & 56 & 5 & \cos(2.7402) \\ 6776 & 121 & 11 & 5 \\ 2464 & 44 & 4 & 2 \end{bmatrix}.$$

Hint. Do not use a calculator. Answer: _____.

(18) Let $A = \begin{bmatrix} 0 & -\frac{1}{2} & 0 & \frac{1}{2} \\ 0 & 0 & \frac{1}{2} & \frac{1}{2} \\ \frac{1}{2} & 0 & -\frac{1}{2} & 0 \\ 1 & 0 & \frac{1}{2} & \frac{1}{2} \end{bmatrix}$. We find A^{-1} using elementary row

operations to convert the 4×8 matrix $\begin{bmatrix} A & \vdots & I_4 \end{bmatrix}$ to the matrix

$\begin{bmatrix} I_4 & \vdots & A^{-1} \end{bmatrix}$.

Give the names of the elementary 4×4 matrices $X_1, \ldots, X_{11}$ that implement the following Gauss-Jordan reduction and fill in the missing matrix entries.

$$\begin{bmatrix} 0 & -\frac{1}{2} & 0 & \frac{1}{2} & \vdots & 1 & 0 & 0 & 0 \\ 0 & 0 & \frac{1}{2} & \frac{1}{2} & \vdots & 0 & 1 & 0 & 0 \\ \frac{1}{2} & 0 & -\frac{1}{2} & 0 & \vdots & 0 & 0 & 1 & 0 \\ 1 & 0 & \frac{1}{2} & \frac{1}{2} & \vdots & 0 & 0 & 0 & 1 \end{bmatrix}$$

$\xrightarrow{X_1} \begin{bmatrix} 1 & 0 & \frac{1}{2} & \frac{1}{2} & \vdots & & & & \\ 0 & 0 & \frac{1}{2} & \frac{1}{2} & \vdots & & & & \\ \frac{1}{2} & 0 & -\frac{1}{2} & 0 & \vdots & & & & \\ 0 & -\frac{1}{2} & 0 & \frac{1}{2} & \vdots & & & & \end{bmatrix}$

$\xrightarrow{X_2} \begin{bmatrix} 1 & 0 & \frac{1}{2} & \frac{1}{2} & \vdots & & & & \\ 0 & 0 & \frac{1}{2} & \frac{1}{2} & \vdots & & & & \\ 0 & 0 & -\frac{3}{4} & -\frac{1}{4} & \vdots & & & & \\ 0 & -\frac{1}{2} & 0 & \frac{1}{2} & \vdots & & & & \end{bmatrix}$

$\xrightarrow{X_3} \begin{bmatrix} 1 & 0 & \frac{1}{2} & \frac{1}{2} & \vdots & & & & \\ 0 & -\frac{1}{2} & 0 & \frac{1}{2} & \vdots & & & & \\ 0 & 0 & -\frac{3}{4} & -\frac{1}{4} & \vdots & & & & \\ 0 & 0 & \frac{1}{2} & \frac{1}{2} & \vdots & & & & \end{bmatrix}$

$$\xrightarrow{X_4}
\begin{bmatrix}
1 & 0 & \frac{1}{2} & \frac{1}{2} & \vdots & \\
0 & 1 & 0 & -1 & \vdots & \\
0 & 0 & -\frac{3}{4} & -\frac{1}{4} & \vdots & \\
0 & 0 & \frac{1}{2} & \frac{1}{2} & \vdots &
\end{bmatrix}$$

$$\xrightarrow{X_5}
\begin{bmatrix}
1 & 0 & \frac{1}{2} & \frac{1}{2} & \vdots \\
0 & 1 & 0 & -1 & \vdots \\
0 & 0 & 1 & \frac{1}{3} & \vdots \\
0 & 0 & \frac{1}{2} & \frac{1}{2} & \vdots
\end{bmatrix}$$

$$\xrightarrow{X_6}
\begin{bmatrix}
1 & 0 & \frac{1}{2} & \frac{1}{2} & \vdots \\
0 & 1 & 0 & -1 & \vdots \\
0 & 0 & 1 & \frac{1}{3} & \vdots \\
0 & 0 & 0 & \frac{1}{3} & \vdots
\end{bmatrix}$$

$$\xrightarrow{X_7}
\begin{bmatrix}
1 & 0 & \frac{1}{2} & \frac{1}{2} & \vdots \\
0 & 1 & 0 & -1 & \vdots \\
0 & 0 & 1 & 0 & \vdots \\
0 & 0 & 0 & \frac{1}{3} & \vdots
\end{bmatrix}$$

$$\xrightarrow{X_8}
\begin{bmatrix}
1 & 0 & \frac{1}{2} & \frac{1}{2} & \vdots \\
0 & 1 & 0 & -1 & \vdots \\
0 & 0 & 1 & 0 & \vdots \\
0 & 0 & 0 & 1 & \vdots
\end{bmatrix}$$

$$\xrightarrow{X_9}
\begin{bmatrix}
1 & 0 & \frac{1}{2} & \frac{1}{2} & \vdots \\
0 & 1 & 0 & 0 & \vdots \\
0 & 0 & 1 & 0 & \vdots \\
0 & 0 & 0 & 1 & \vdots
\end{bmatrix}$$

$$\xrightarrow{X_{10}}
\begin{bmatrix}
1 & 0 & \frac{1}{2} & 0 & \vdots \\
0 & 1 & 0 & 0 & \vdots \\
0 & 0 & 1 & 0 & \vdots \\
0 & 0 & 0 & 1 & \vdots
\end{bmatrix}$$

$$\xrightarrow{X_{11}}
\begin{bmatrix}
1 & 0 & 0 & 0 & \vdots \\
0 & 1 & 0 & 0 & \vdots \\
0 & 0 & 1 & 0 & \vdots \\
0 & 0 & 0 & 1 & \vdots
\end{bmatrix}$$

Answer: $X_1 =$ _____, $X_2 =$ _____, $X_3 =$ _____, $X_4 =$ _____, $X_5 =$ _____, $X_6 =$ _____, $X_7 =$ _____, $X_8 =$ _____. $X_9 =$ _____, $X_{10} =$ _____, $X_{11} =$ _____.

(19) The matrix

$$H = \begin{bmatrix}
1 & \frac{1}{2} & \frac{1}{3} & \frac{1}{4} \\
\frac{1}{2} & \frac{1}{3} & \frac{1}{4} & \frac{1}{5} \\
\frac{1}{3} & \frac{1}{4} & \frac{1}{5} & \frac{1}{6} \\
\frac{1}{4} & \frac{1}{5} & \frac{1}{6} & \frac{1}{7}
\end{bmatrix}$$

is the 4×4 HILBERT MATRIX. Use Gauss-Jordan reduction to compute $K = H^{-1}$. Then K_{44} is (exactly) _____. Now, create a new matrix H' by replacing each entry in H by its approximation to 3 decimal places. (For example, replace $\frac{1}{6}$ by 0.167.)

Use Gauss-Jordan reduction again to find the inverse K' of H'. Then K'_{44} is _____.

(20) Suppose that A is a square matrix with determinant 7. Then

(a) $\det(P_{24}A) = $ _____.
(b) $\det(E_{23}(-4)A) = $ _____.
(c) $\det(M_3(2)A) = $ _____.

2.3 Problems

(1) For this problem assume that we know the following: If X is an $m \times m$ matrix, if Y is an $m \times n$ matrix and if $\mathbf{0}$ and $\mathbf{I}$ are zero and identity matrices of appropriate sizes, then $\det \begin{bmatrix} X & Y \\ \mathbf{0} & \mathbf{I} \end{bmatrix} = \det X$.

Let A be an $m \times n$ matrix and B be an $n \times m$ matrix. Prove carefully that

$$\det \begin{bmatrix} \mathbf{0} & A \\ -B & \mathbf{I} \end{bmatrix} = \det AB.$$

Hint. Consider the product $\begin{bmatrix} \mathbf{0} & A \\ -B & \mathbf{I} \end{bmatrix} \begin{bmatrix} \mathbf{I} & \mathbf{0} \\ B & \mathbf{I} \end{bmatrix}$.

(2) Let A and B be $n \times n$-matrices. Your good friend Fred R. Dimm believes that

$$\det \begin{bmatrix} A & B \\ B & A \end{bmatrix} = \det(A + B) \det(A - B).$$

He offers the following argument to support this claim:

$$\det \begin{bmatrix} A & B \\ B & A \end{bmatrix} = \det(A^2 - B^2)$$
$$= \det[(A + B)(A - B)]$$
$$= \det(A + B) \det(A - B).$$

(a) Comment (helpfully) on his "proof". In particular, explain carefully why each of the three steps in his "proof" is correct or incorrect. (That is, provide a proof or a counterexample to each step.)

(b) Is the result he is trying to prove actually true?

Hint: Consider the product $\begin{bmatrix} I & B \\ 0 & A - B \end{bmatrix} \begin{bmatrix} A + B & 0 \\ 0 & I \end{bmatrix}$.

(3) Let x be a fixed real number that is not an integer multiple of π. For each natural number n let $A_n = [a_{jk}]$ be the $n \times n$-matrix defined by

$$a_{jk} = \begin{cases} 0, & \text{for } |j - k| > 1 \\ 1, & \text{for } |j - k| = 1 \\ 2\cos x, & \text{for } j = k. \end{cases}$$

Show that $\det A_n = \dfrac{\sin(n + 1)x}{\sin x}$. *Hint.* For each integer n let $D_n = \det A_n$ and prove that

$$D_{n+2} - 2D_{n+1}\cos x + D_n = 0.$$

(Use mathematical induction.)

2.4 Answers to Odd-Numbered Exercises

(1) $\begin{bmatrix} 1 & 0 & 0 & 0 \\ 0 & 1 & 0 & 0 \\ 0 & 0 & 4 & 0 \\ 0 & 0 & 0 & 1 \end{bmatrix}$

(3) $\begin{bmatrix} 1 & 0 & 0 & 0 \\ 0 & 1 & -2 & 0 \\ 0 & 0 & 1 & 0 \\ 0 & 0 & 0 & 1 \end{bmatrix}$

(5) $\begin{bmatrix} 1 & 0 & 0 & 0 \\ 0 & 0 & 0 & 1 \\ 0 & 0 & 1 & 0 \\ 0 & 1 & 0 & 0 \end{bmatrix}, \begin{bmatrix} 1 & 0 & 0 & 0 \\ 0 & 1 & 0 & 0 \\ 0 & 0 & 1 & 0 \\ 0 & 0 & 0 & 1 \end{bmatrix}$

(7) $-8, -1$

(9) $P_{12}, E_{31}(1), E_{32}(-1), M_3(-\frac{1}{9}), E_{23}(-3), M_2(\frac{1}{2}), E_{13}(1), E_{12}(-1)$

(11) (a) $3, 2, 3, -2$

(b) $\begin{bmatrix} 1 & -1 & 1 \\ -2 & \frac{1}{2} & 0 \\ 1 & 0 & 0 \end{bmatrix}$

(13) 100, 0, −72

(15) −10

(17) 6

(19) 2800, −1329.909

Chapter 3

VECTOR GEOMETRY IN $\mathbb{R}^n$

3.1 Background

Topics: inner (dot) products, cross products, lines and planes in 3-space, norm of a vector, angle between vectors, the *law of sines.*

Notation 3.1.1. There are many more or less standard notations for the inner product (or dot product) of two vectors $\mathbf{x}$ and $\mathbf{y}$. The two that we will use interchangeably in these exercises are $\mathbf{x} \cdot \mathbf{y}$ and $\langle \mathbf{x}, \mathbf{y} \rangle$.

Definition 3.1.2. If $\mathbf{x}$ is a vector in $\mathbb{R}^n$, then the NORM (or LENGTH) of $\mathbf{x}$ is defined by

$$\|\mathbf{x}\| = \sqrt{\langle \mathbf{x}, \mathbf{x} \rangle}.$$

Definition 3.1.3. Let $\mathbf{x}$ and $\mathbf{y}$ be nonzero vectors in $\mathbb{R}^n$. Then $\measuredangle(\mathbf{x}, \mathbf{y})$, the ANGLE between $\mathbf{x}$ and $\mathbf{y}$, is defined by

$$\measuredangle(\mathbf{x}, \mathbf{y}) = \arccos \frac{\langle \mathbf{x}, \mathbf{y} \rangle}{\|\mathbf{x}\| \, \|\mathbf{y}\|}.$$

Theorem 3.1.4 (Cauchy-Schwarz inequality). *If* $\mathbf{x}$ *and* $\mathbf{y}$ *are vectors in* $\mathbb{R}^n$*, then*

$$|\langle \mathbf{x}, \mathbf{y} \rangle| \leq \|\mathbf{x}\| \, \|\mathbf{y}\|.$$

(We will often refer to this just as the *Schwarz inequality.*)

Definition 3.1.5. If $\mathbf{x} = (x_1, x_2, x_3)$ and $\mathbf{y} = (y_1, y_2, y_3)$ are vectors in $\mathbb{R}^3$, then their CROSS PRODUCT, denoted by $\mathbf{x} \times \mathbf{y}$, is the vector $(x_2 y_3 - x_3 y_2, x_3 y_1 - x_1 y_3, x_1 y_2 - x_2 y_1)$.

3.2   Exercises

(1) The angle between the vectors $(1, 0, -1, 3)$ and $(1, \sqrt{3}, 3, -3)$ in $\mathbb{R}^4$ is $a\pi$ where $a =$ _____.

(2) Find the angle θ between the vectors $\mathbf{x} = (3, -1, 1, 0, 2, 1)$ and $\mathbf{y} = (2, -1, 0, \sqrt{2}, 2, 1)$ in $\mathbb{R}^6$. Answer: $\theta =$ _____.

(3) If $a_1, \ldots, a_n > 0$, then

$$\left(\sum_{j=1}^{n} a_j \right) \left(\sum_{k=1}^{n} \frac{1}{a_k} \right) \geq n^2.$$

The proof of this is obvious from the *Cauchy-Schwarz inequality* when we choose the vectors $\mathbf{x}$ and $\mathbf{y}$ as follows:

$\mathbf{x} =$ _____ and $\mathbf{y} =$ _____.

(4) Find all real numbers α such that the angle between the vectors $2\mathbf{i} + 2\mathbf{j} + (\alpha - 2)\mathbf{k}$ and $2\mathbf{i} + (\alpha - 2)\mathbf{j} + 2\mathbf{k}$ is $\frac{\pi}{3}$. Answer: $\alpha =$ _____ and _____.

(5) Which of the angles (if any) of triangle ABC, with $A = (1, -2, 0)$, $B = (2, 1, -2)$, and $C = (6, -1, -3)$, is a right angle? Answer: the angle at vertex _____.

(6) The hydrogen atoms of a methane molecule (CH_4) are located at $(0, 0, 0)$, $(1, 1, 0)$, $(0, 1, 1)$, and $(1, 0, 1)$ while the carbon atom is at $(\frac{1}{2}, \frac{1}{2}, \frac{1}{2})$. Find the cosine of the angle θ between two rays starting at the carbon atom and going to different hydrogen atoms.

Answer: $\cos \theta =$ _____.

(7) If $a, b, c, d, e, f \in \mathbb{R}$, then

$$|ad + be + cf| \leq \sqrt{a^2 + b^2 + c^2} \sqrt{d^2 + e^2 + f^2}.$$

The proof of this inequality is obvious since this is just the *Cauchy-Schwarz inequality* where $x = ($ _____, _____, _____ $)$ and $y = ($ _____, _____, _____ $)$.

(8) The volume of the parallelepiped generated by the three vectors $\mathbf{i} + 2\mathbf{j} - \mathbf{k}$, $\mathbf{j} + \mathbf{k}$, and $3\mathbf{i} - \mathbf{j} + 2\mathbf{k}$ is _____.

(9) The equations of the line containing the points $(3, -1, 4)$ and $(7, 9, 10)$ are
$$\frac{x - 3}{2} = \frac{y - j}{b} = \frac{z - k}{c}$$
where $b =$ _____, $c =$ _____, $j =$ _____, and $k =$ _____.

(10) The equations of the line containing the points $(5, 2, -1)$ and $(9, -4, 1)$ are
$$\frac{x - h}{a} = \frac{y - 2}{-3} = \frac{z - k}{c}$$
where $a =$ _____, $c =$ _____, $h =$ _____, and $k =$ _____.

(11) Find the equations of the line containing the point $(1, 0, -1)$ that is parallel to the line $\dfrac{x - 4}{2} = \dfrac{2y - 3}{5} = \dfrac{3z - 7}{6}$.

Answer: $\dfrac{x - h}{a} = \dfrac{y - j}{b} = \dfrac{z + 1}{4}$ where $a =$ _____, $b =$ _____, $h =$ _____, and $j =$ _____.

(12) The equation of the plane containing the points $(0, -1, 1)$, $(1, 0, 2)$, and $(3, 0, 1)$ is $x + by + cz = d$ where $b =$ _____, $c =$ _____, and $d =$ _____.

(13) The equation of the plane which passes through the points $(0, -1, -1)$, $(5, 0, 1)$, and $(4, -1, 0)$ is $ax + by + cz = 1$ where $a =$ _____, $b =$ _____, and $c =$ _____.

(14) The angle between the planes $4x + 4z - 16 = 0$ and $-2x + 2y - 13 = 0$ is $\dfrac{a}{b}\pi$ where $a =$ _____ and $b =$ _____.

(15) Suppose that $\mathbf{u} \in \mathbb{R}^3$ is a vector that lies in the first quadrant of the xy-plane and has length 3 and that $\mathbf{v} \in \mathbb{R}^3$ is a vector that lies along the positive z-axis and has length 5. Then
   (a) $\|\mathbf{u} \times \mathbf{v}\| =$ _____;
   (b) the x-coordinate of $\mathbf{u} \times \mathbf{v}$ is _____ 0 (choose $<$, $>$, or $=$);
   (c) the y-coordinate of $\mathbf{u} \times \mathbf{v}$ is _____ 0 (choose $<$, $>$, or $=$); and
   (d) the z-coordinate of $\mathbf{u} \times \mathbf{v}$ is _____ 0 (choose $<$, $>$, or $=$).

(16) Suppose that $\mathbf{u}$ and $\mathbf{v}$ are vectors in a vector space both of length $2\sqrt{2}$ and that the length of $\mathbf{u} - \mathbf{v}$ is also $2\sqrt{2}$. Then $\|\mathbf{u} + \mathbf{v}\| =$ _____ and the angle between $\mathbf{u}$ and $\mathbf{v}$ is _____.

(17) If $\mathbf{a} = 3\mathbf{i} + 4\mathbf{j} + 12\mathbf{k}$, $\mathbf{b} = 3\mathbf{i} + 4\mathbf{j} - 12\mathbf{k}$, and $\mathbf{c} = \frac{1}{8}\mathbf{i} - \frac{1}{12}\mathbf{j} + \frac{1}{16}\mathbf{k}$, then $\langle \mathbf{c}, \mathbf{a} \times \mathbf{b} \rangle =$ _____.

3.3   Problems

(1) Show that if a, b, $c > 0$, then $\left(\frac{1}{2}a + \frac{1}{3}b + \frac{1}{6}c\right)^2 \le \frac{1}{2}a^2 + \frac{1}{3}b^2 + \frac{1}{6}c^2$.

(2) Show that if $a_1, \ldots, a_n$, $w_1, \ldots, w_n > 0$ and $\sum_{k=1}^{n} w_k = 1$, then

$$\left(\sum_{k=1}^{n} a_k w_k\right)^2 \le \sum_{k=1}^{n} a_k{}^2 w_k.$$

(3) Prove that if $(a_1, a_2, \ldots)$ is a sequence of real numbers such that the series $\sum_{k=1}^{\infty} a_k{}^2$ converges, then the series $\sum_{k=1}^{\infty} \frac{1}{k} a_k$ converges absolutely.

You may find the following steps helpful in organizing your solution.

   (i) First of all, make sure that you recall the difference between a sequence of numbers $(c_1, c_2, \ldots)$ and an infinite series $\sum_{k=1}^{\infty} c_k$.

   (ii) The key to this problem is an important theorem from third term Calculus:

> *A nondecreasing sequence of real numbers converges if and only if it is bounded.* $\qquad (*)$

   (Make sure that you know the meanings of all the terms used here.)

   (iii) The hypothesis of the result we are trying to prove is that the series $\sum_{k=1}^{\infty} a_k{}^2$ converges. What, exactly, does this mean?

   (iv) For each natural number n let $b_n = \sum_{k=1}^{n} a_k{}^2$. Rephrase (iii) in terms of the sequence (b_n).

   (v) Is the sequence (b_n) nondecreasing?

   (vi) What, then, does $(*)$ say about the sequence (b_n)?

   (vii) For each natural number n let $c_n = \sum_{k=1}^{n} \frac{1}{k^2}$. What do we know about the sequence (c_n) from third term Calculus? What does $(*)$ say about the sequence (c_n)?

   (viii) The conclusion we are trying to prove is that the series $\sum_{k=1}^{\infty} \frac{1}{k} a_k$ converges absolutely. What does this mean?

(ix) For each natural number n let $s_n = \sum\limits_{k=1}^{n} \dfrac{1}{k}|a_k|$. Rephrase (viii) in terms of the sequence (s_n).

(x) Explain how for each n we may regard the number s_n as the dot product of two vectors in $\mathbb{R}^n$.

(xi) Apply the *Cauchy-Schwarz inequality* to the dot product in (x). Use (vi) and (vii) to establish that the sequence (s_n) is bounded above.

(xii) Use $(*)$ one last time — keeping in mind what you said in (ix).

(4) Let $\mathbf{a}$ and $\mathbf{b}$ be vectors in $\mathbb{R}^3$. Without using the components of $\mathbf{a}$ and $\mathbf{b}$ prove that $\|\mathbf{a} \times \mathbf{b}\|^2 = \|\mathbf{a}\|^2\|\mathbf{b}\|^2 - \langle \mathbf{a}, \mathbf{b} \rangle^2$.

(5) Let $\mathbf{a}$, $\mathbf{b}$, and $\mathbf{c}$ be vectors in $\mathbb{R}^3$.

(a) Prove that if $\mathbf{a} + \mathbf{b} + \mathbf{c} = \mathbf{0}$, then $\mathbf{a} \times \mathbf{b} = \mathbf{b} \times \mathbf{c} = \mathbf{c} \times \mathbf{a}$.

(b) Use part (a) to prove the *law of sines*.

3.4   Answers to Odd-Numbered Exercises

(1) $\frac{3}{4}$

(3) $\left(\sqrt{a_1}, \sqrt{a_2}, \ldots, \sqrt{a_n}\right)$, $\left(\frac{1}{\sqrt{a_1}}, \frac{1}{\sqrt{a_2}}, \ldots, \frac{1}{\sqrt{a_n}}\right)$

(5) B

(7) a, b, c, d, e, f

(9) $5, 3, -1, 4$

(11) $4, 5, 1, 0$

(13) $1, 3, -4$

(15) (a) 15
     (b) $>$
     (c) $<$
     (d) $=$

(17) -18

Part 2

VECTOR SPACES

Chapter 4

VECTOR SPACES

4.1 Background

Topics: real and complex vector spaces, vectors, scalars.

CAUTION. In the following definition $\mathbb{F}$ may be taken to be an arbitrary field. In this study guide, however, we will deal exclusively with two cases only, $\mathbb{F} = \mathbb{R}$ (the field of real numbers) and $\mathbb{F} = \mathbb{C}$ (the field of complex numbers, which, if you are not already familiar with it, is introduced in Chapter 16).

Definition 4.1.1. A VECTOR SPACE is a set V together with operations of addition and scalar multiplication that satisfy the following axioms:

(1) if $\mathbf{x}, \mathbf{y} \in V$, then $\mathbf{x} + \mathbf{y} \in V$;

(2) $(\mathbf{x} + \mathbf{y}) + \mathbf{z} = \mathbf{x} + (\mathbf{y} + \mathbf{z})$ for every $\mathbf{x}, \mathbf{y}, \mathbf{z} \in V$ (associativity);

(3) there exists $\mathbf{0} \in V$ such that $\mathbf{x} + \mathbf{0} = \mathbf{x}$ for every $\mathbf{x} \in V$ (existence of additive identity);

(4) for every $\mathbf{x} \in V$ there exists $-\mathbf{x} \in V$ such that $\mathbf{x} + (-\mathbf{x}) = \mathbf{0}$ (existence of additive inverses);

(5) $\mathbf{x} + \mathbf{y} = \mathbf{y} + \mathbf{x}$ for every $\mathbf{x}, \mathbf{y} \in V$ (commutativity);

(6) if $\alpha \in \mathbb{F}$ and $\mathbf{x} \in V$, then $\alpha\mathbf{x} \in V$;

(7) $\alpha(\mathbf{x} + \mathbf{y}) = \alpha\mathbf{x} + \alpha\mathbf{y}$ for every $\alpha \in \mathbb{F}$ and every $\mathbf{x}, \mathbf{y} \in V$;

(8) $(\alpha + \beta)\mathbf{x} = \alpha\mathbf{x} + \beta\mathbf{x}$ for every $\alpha, \beta \in \mathbb{F}$ and every $\mathbf{x} \in V$;

(9) $(\alpha\beta)\mathbf{x} = \alpha(\beta\mathbf{x})$ for every $\alpha, \beta \in \mathbb{F}$ and every $\mathbf{x} \in V$; and

(10) $1\mathbf{x} = \mathbf{x}$ for every $\mathbf{x} \in V$.

When $\mathbb{F} = \mathbb{R}$ we speak of V as a REAL VECTOR SPACE and when $\mathbb{F} = \mathbb{C}$ we call it a COMPLEX VECTOR SPACE.

4.2   Exercises

(1) Let V be the set of all real numbers. Define an operation of "addition" by

$$x \boxplus y = \text{ the maximum of } x \text{ and } y$$

for all $x, y \in V$. Define an operation of "scalar multiplication" by

$$\alpha \boxdot x = \alpha x$$

for all $\alpha \in \mathbb{R}$ and $x \in V$.

Under the operations $\boxplus$ and $\boxdot$ the set V is not a vector space. The vector space axioms (see 4.1.1 (1)–(10)) that *fail* to hold are _____, _____, _____, and _____.

(2) Let V be the set of all real numbers x such that $x \geq 0$. Define an operation of "addition" by

$$x \boxplus y = xy + 1$$

for all $x, y \in V$. Define an operation of "scalar multiplication" by

$$\alpha \boxdot x = \alpha^2 x$$

for all $\alpha \in \mathbb{R}$ and $x \in V$.

Under the operations $\boxplus$ and $\boxdot$ the set V _____ (is/is not) a vector space. If it is not, list all the vector space axioms (see 4.1.1 (1)–(10)) that *fail* to hold. Answer: The axioms that are *not* satisfied are _____.

(3) Let V be $\mathbb{R}^2$, the set of all ordered pairs (x, y) of real numbers. Define an operation of "addition" by

$$(u, v) \boxplus (x, y) = (u + x + 1, v + y + 1)$$

for all (u, v) and (x, y) in V. Define an operation of "scalar multiplication" by

$$\alpha \boxdot (x, y) = (\alpha x, \alpha y)$$

for all $\alpha \in \mathbb{R}$ and $(x, y) \in V$.

Under the operations $\boxplus$ and $\boxdot$ the set V is not a vector space. The vector space axioms (see 4.1.1 (1)–(10)) that *fail* to hold are _____ and _____.

(4) Let V be $\mathbb{R}^2$, the set of all ordered pairs (x, y) of real numbers. Define an operation of "addition" by

$$(u, v) \boxplus (x, y) = (u + x, 0)$$

for all (u, v) and (x, y) in V. Define an operation of "scalar multiplication" by

$$\alpha \boxdot (x, y) = (\alpha x, \alpha y)$$

for all $\alpha \in \mathbb{R}$ and $(x, y) \in V$.

Under the operations $\boxplus$ and $\boxdot$ the set V is not a vector space. The vector space axioms (see 4.1.1 (1)–(10)) that *fail* to hold are _____, _____, and _____.

(5) Let V be the set of all $n \times n$ matrices of real numbers. Define an operation of "addition" by

$$A \boxplus B = \frac{1}{2}(AB + BA)$$

for all A, $B \in V$. Define an operation of "scalar multiplication" by

$$\alpha \boxdot A = \mathbf{0}$$

for all $\alpha \in \mathbb{R}$ and $A \in V$.

Under the operations $\boxplus$ and $\boxdot$ the set V is not a vector space. The vector space axioms (see 4.1.1 (1)–(10)) that *fail* to hold are _____, _____, and _____.

(6) Below are portions of proofs of four results about vector spaces which establish the fact that multiplying a vector x by the scalar -1 produces $-x$, the additive inverse of x. Fill in the missing steps and the missing reasons. Choose reasons from the following list.

$$
\begin{array}{ll}
\text{(H)} & \text{Hypothesis} \\
\text{(1)–(10)} & \text{Vector space axioms, see 4.1.1} \\
\text{(PA)} & \text{Proposition A} \\
\text{(PB)} & \text{Proposition B} \\
\text{(PC)} & \text{Proposition C} \\
\text{(RN)} & \text{Property of the Real Numbers}
\end{array}
$$

Proposition 4.2.1 (A). *A vector x in a vector space V has at most one additive inverse. That is, if $\mathbf{y}$ and $\mathbf{z}$ are vectors in V such that $\mathbf{x} + \mathbf{y} = \mathbf{0}$ and $\mathbf{x} + \mathbf{z} = \mathbf{0}$, then $\mathbf{y} = \mathbf{z}$.*

Proof. Suppose that $x + y = 0$ and $x + z = 0$. Then

$$
\begin{aligned}
y &= \underline{\hspace{3cm}} &&\text{(reason: \underline{\hspace{1.5cm}})} \\
&= y + (x + z) &&\text{(reason: \underline{\hspace{1.5cm}})} \\
&= \underline{\hspace{3cm}} &&\text{(reason: (2))} \\
&= (x + y) + z &&\text{(reason: \underline{\hspace{1.5cm}})} \\
&= \underline{\hspace{3cm}} &&\text{(reason: (H))} \\
&= \underline{\hspace{3cm}} &&\text{(reason: (5))} \\
&= z &&\text{(reason: \underline{\hspace{1.5cm}}).} \qquad \square
\end{aligned}
$$

Proposition 4.2.2 (B). *If $x \in V$ where V is a vector space and $x + x = x$, then $x = 0$.*

Proof. If $x \in V$ and $x + x = x$, then

$$
\begin{aligned}
x &= x + 0 &&\text{(reason: \underline{\hspace{1.5cm}})} \\
&= \underline{\hspace{3cm}} &&\text{(reason: (4))} \\
&= (x + x) + (-x) &&\text{(reason: \underline{\hspace{1.5cm}})} \\
&= \underline{\hspace{3cm}} &&\text{(reason: (H))} \\
&= 0 &&\text{(reason: \underline{\hspace{1.5cm}}).} \qquad \square
\end{aligned}
$$

Proposition 4.2.3 (C). *If x is a vector in a vector space V, then $0x = 0$.*

Proof. If $x \in V$, then

$$
\begin{aligned}
0x &= (0 + 0)x &&\text{(reason: \underline{\hspace{1.5cm}})} \\
&= \underline{\hspace{2.5cm}} &&\text{(reason: (8))}
\end{aligned}
$$

Thus $0x = 0$ (reason: \underline{\hspace{2.5cm}}). $\square$

Proposition 4.2.4 (D). *If x is a vector in a vector space V, then $(-1)x$ is $-x$, the additive inverse of x.*

Proof. If $x \in V$, then

$$
\begin{aligned}
x + (-1) \cdot x &= \underline{\hspace{5cm}} &&\text{(reason: (10))} \\
&= (1 + (-1)) \cdot x &&\text{(reason: \underline{\hspace{1.5cm}})} \\
&= 0 \cdot x &&\text{(reason: \underline{\hspace{1.5cm}})} \\
&= 0 &&\text{(reason: \underline{\hspace{1.5cm}}).}
\end{aligned}
$$

It then follows immediately from \underline{\hspace{2.5cm}} that $(-1) \cdot x = -x$. $\square$

(7) In this exercise we prove that multiplying the zero vector by an arbitrary scalar produces the zero vector. For each step of the proof give the appropriate reason. Choose reasons from the following list.

$$(1)-(10) \quad \text{Vector space axioms 4.1.1.}$$
$$(PB) \quad \text{Proposition 4.2.2}$$
$$(RN) \quad \text{Property of the Real Numbers}$$

Proposition 4.2.5 (E). *If* $\mathbf{0}$ *is the zero vector in a vector space and* α *is a scalar, then* $\alpha \cdot \mathbf{0} = \mathbf{0}$.

Proof. Let $\mathbf{0}$ be the zero vector of some vector space. Then for every scalar α

$$\alpha \cdot \mathbf{0} = \alpha \cdot (\mathbf{0} + \mathbf{0}) \quad \text{(reason: _____)}$$
$$= \alpha \cdot \mathbf{0} + \alpha \cdot \mathbf{0} \quad \text{(reason: _____)}$$

It then follows immediately from _____ that $\alpha \cdot \mathbf{0} = \mathbf{0}$. □

(8) In this exercise we prove that the product of a scalar and a vector is zero if and only if either the scalar or the vector is zero. After each step of the proof give the appropriate reason. Choose reasons from the following list.

$$(H) \quad \text{Hypothesis.}$$
$$(1)-(10) \quad \text{Vector space axioms 4.1.1.}$$
$$(PC), (PE) \quad \text{Propositions 4.2.3 and 4.2.5, respectively.}$$
$$(RN) \quad \text{Property of the Real Numbers.}$$

Proposition 4.2.6. *Suppose that* $\mathbf{x}$ *is a vector and* α *is a scalar. Then* $\alpha \mathbf{x} = \mathbf{0}$ *if and only if* $\alpha = 0$ *or* $\mathbf{x} = \mathbf{0}$.

Proof. We have already shown in _____ and _____ that if $\alpha = 0$ or $\mathbf{x} = \mathbf{0}$, then $\alpha \mathbf{x} = \mathbf{0}$.

To prove the converse we suppose that $\alpha \mathbf{x} = \mathbf{0}$ and that $\alpha \neq 0$; and we prove that $\mathbf{x} = \mathbf{0}$. This conclusion results from the following easy calculation:

$$\mathbf{x} = 1 \cdot \mathbf{x} \quad \text{(reason: _____)}$$
$$= \left(\frac{1}{\alpha} \cdot \alpha\right) \cdot \mathbf{x} \quad \text{(reasons: _____ and _____)}$$
$$= \frac{1}{\alpha} \cdot (\alpha \cdot \mathbf{x}) \quad \text{(reason: _____)}$$
$$= \frac{1}{\alpha} \cdot \mathbf{0} \quad \text{(reason: _____)}$$
$$= \mathbf{0} \quad \text{(reason: _____).} \qquad\qquad \square$$

4.3   Problems

(1) Prove that if V is a vector space, then its additive identity is unique. That is, show that if $\mathbf{0}$ and $\widetilde{\mathbf{0}}$ are vectors in V such that $\mathbf{x} + \mathbf{0} = \mathbf{x}$ for all $x \in V$ and $\mathbf{x} + \widetilde{\mathbf{0}} = \mathbf{x}$ for all $x \in V$, then $\mathbf{0} = \widetilde{\mathbf{0}}$.

(2) Let V be the set of all real numbers x such that $x > 0$. Define an operation of "addition" by

$$x \boxplus y = xy$$

for all $x, y \in V$. Define an operation of "scalar multiplication" by

$$\alpha \boxdot x = x^{\alpha}$$

for all $\alpha \in \mathbb{R}$ and $x \in V$.

Prove that under the operations $\boxplus$ and $\boxdot$ the set V is a vector space.

(3) With the usual operations of addition and scalar multiplication the set of all $n \times n$ matrices of real numbers is a vector space: in particular, all the vector space axioms (see 4.1.1 (1)–(10)) are satisfied. Explain clearly why the set of all nonsingular $n \times n$ matrices of real numbers is *not* a vector space under these same operations.

4.4   Answers to Odd-Numbered Exercises

(1) 3, 4, 7, 8

(3) 7, 8

(5) 2, 4, 10

(7) 3, 7, PB

Chapter 5

SUBSPACES

5.1 Background

Topics: subspaces of a vector space

Definition 5.1.1. A nonempty subset of M of a vector space V is a SUBSPACE of V if it is closed under addition and scalar multiplication. (That is: if $\mathbf{x}$ and $\mathbf{y}$ belong to M, so does $\mathbf{x} + \mathbf{y}$; and if $\mathbf{x}$ belongs to M and $\alpha \in \mathbb{R}$, then $\alpha \mathbf{x}$ belongs to M.)

Notation 5.1.2. We use the notation $M \preceq V$ to indicate that M is a subspace of a vector space V.

Notation 5.1.3. Here are some frequently encountered families of functions:

$$\mathcal{F} = \mathcal{F}[a, b] = \{f : f \text{ is a real valued function on the interval } [a, b]\} \tag{5.1.1}$$

$$\mathcal{P} = \mathcal{P}[a, b] = \{p : p \text{ is a polynomial function on } [a, b]\} \tag{5.1.2}$$

$$\mathcal{P}_4 = \mathcal{P}_4[a, b] = \{p \in \mathcal{P} : \text{ the degree of } p \text{ is less than } 4\} \tag{5.1.3}$$

$$\mathcal{Q}_4 = \mathcal{Q}_4[a, b] = \{p \in \mathcal{P} : \text{ the degree of } p \text{ is equal to } 4\} \tag{5.1.4}$$

$$\mathcal{C} = \mathcal{C}[a, b] = \{f \in \mathcal{F} : f \text{ is continuous}\} \tag{5.1.5}$$

$$\mathcal{D} = \mathcal{D}[a, b] = \{f \in \mathcal{F} : f \text{ is differentiable}\} \tag{5.1.6}$$

$$\mathcal{K} = \mathcal{K}[a, b] = \{f \in \mathcal{F} : f \text{ is a constant function}\} \tag{5.1.7}$$

$$\mathcal{B} = \mathcal{B}[a, b] = \{ f \in \mathcal{F} \colon f \text{ is bounded} \} \qquad (5.1.8)$$

$$\mathcal{J} = \mathcal{J}[a, b] = \{ f \in \mathcal{F} \colon f \text{ is integrable} \} \qquad (5.1.9)$$

(A function $f \in \mathcal{F}$ is BOUNDED if there exists a number $M \geq 0$ such that $|f(x)| \leq M$ for all x in $[a, b]$. It is (RIEMANN) INTEGRABLE if it is bounded and $\int_a^b f(x)\, dx$ exists.)

Definition 5.1.4. If A and B are subsets of a vector space then the SUM of A and B, denoted by $A + B$, is defined by

$$A + B := \{ \mathbf{a} + \mathbf{b} \colon \mathbf{a} \in A \text{ and } \mathbf{b} \in B \}.$$

Definition 5.1.5. Let M and N be subspaces of a vector space V. If $M \cap N = \{\mathbf{0}\}$ and $M + N = V$, then V is the (INTERNAL) DIRECT SUM of M and N. In this case we write

$$V = M \oplus N.$$

In this case the subspaces M and N are COMPLEMENTARY and each is the COMPLEMENT of the other.

Definition 5.1.6. Let V and W be vector spaces. If addition and scalar multiplication are defined on the Cartesian product $V \times W$ by

$$(v, w) + (x, y) := (v + x, w + y)$$

and

$$\alpha(v, w) := (\alpha v, \alpha w)$$

for all v, $x \in V$, all w, $y \in W$, and all $\alpha \in \mathbb{F}$, then $V \times W$ becomes a vector space. (This is called the PRODUCT or (EXTERNAL) DIRECT SUM of V and W. It is frequently denoted by $V \oplus W$.)

Notice that the same notation is used for internal and external direct sums. Some authors make a notational distinction by using $\oplus_i$ and $\oplus_e$ (or $\oplus$ and $\boxplus$) to denote these two concepts. Others regard it as somewhat pedantic on the grounds that context should make it clear whether or not the summands are currently being regarded as subspaces of some larger vector space. The matter becomes even worse when, in inner product spaces, we encounter 'orthogonal direct sums' (see Definition 18.1.6 in Chapter 18).

Definition 5.1.7. Let M be a subspace of a vector space V. Define an equivalence relation $\sim$ on V by

$$x \sim y \quad \text{if and only if} \quad y - x \in M.$$

For each $x \in V$ let $[x]$ be the equivalence class containing x. Let V/M be the set of all equivalence classes of elements of V. For $[x]$ and $[y]$ in V/M define

$$[x] + [y] := [x + y]$$

and for $\alpha \in \mathbb{R}$ and $[x] \in V/M$ define

$$\alpha[x] := [\alpha x].$$

Under these operations V/M becomes a vector space. It is the QUOTIENT SPACE of V by M. The notation V/M is usually read "V mod M".

Note that the preceding "definition" contains many statements of fact. These must be verified. (See Problem 6.)

5.2 Exercises

(1) One of the following is a subspace of $\mathbb{R}^3$. Which one?

The set of points (x, y, z) in $\mathbb{R}^3$ such that

(a) $x + 2y - 3z = 4$.
(b) $\frac{x-1}{2} = \frac{y+2}{3} = \frac{z}{4}$.
(c) $x + y + z = 0$ and $x - y + z = 1$.
(d) $x = -z$ and $x = z$.
(e) $x^2 + y^2 = z$.
(f) $\frac{x}{2} = \frac{y-3}{5}$.

Answer: (___) is a subspace of $\mathbb{R}^3$.

(2) The smallest subspace of $\mathbb{R}^3$ containing the vectors $(2, -3, -3)$ and $(0, 3, 2)$ is the plane whose equation is $ax + by + 6z = 0$ where $a = $ _____, and $b = $ _____.

(3) The smallest subspace of $\mathbb{R}^3$ containing the vectors $(0, -3, 6)$ and $(0, 1, -2)$ is the line whose equations are $x = a$ and $z = by$ where $a = $ _____, and $b = $ _____.

(4) Let $\mathbb{R}^\infty$ denote the vector space of all sequences of real numbers. (Addition and scalar multiplication are defined coordinatewise.) In each of the following a subset of $\mathbb{R}^\infty$ is described. Write *yes* if the set is a subspace of $\mathbb{R}^\infty$ and *no* if it is not.

   (a) Sequences that have infinitely many zeros (for example, $(1, 1, 0, 1, 1, 0, 1, 1, 0, \ldots)$). Answer: _____.

   (b) Sequences that are eventually zero. (A sequence (x_k) is *eventually zero* if there is an index n_0 such that $x_n = 0$ whenever $n \geq n_0$.) Answer: _____.

   (c) Sequences that are absolutely summable. (A sequence (x_k) is *absolutely summable* if $\sum_{k=1}^{\infty} |x_k| < \infty$.) Answer: _____.

   (d) Bounded sequences. (A sequence (x_k) is *bounded* if there is a positive number M such that $|x_k| \leq M$ for every k.) Answer: _____.

   (e) Decreasing sequences. (A sequence (x_k) is *decreasing* if $x_{n+1} \leq x_n$ for each n.) Answer: _____.

   (f) Convergent sequences. Answer: _____.

   (g) Arithmetic progressions. (A sequence (x_k) is *arithmetic* if it is of the form $(a, a + k, a + 2k, a + 3k, \ldots)$ for some constant k.) Answer: _____.

   (h) Geometric progressions. (A sequence (x_k) is *geometric* if it is of the form $(a, ka, k^2 a, k^3 a, \ldots)$ for some constant k.) Answer: _____.

(5) Let M and N be subspaces of a vector space V. Consider the following subsets of V.

   (a) $M \cap N$. (A vector $\mathbf{v}$ belongs to $M \cap N$ if it belongs to both M and N.)
   (b) $M \cup N$. (A vector $\mathbf{v}$ belongs to $M \cup N$ if it belongs to either M or N.)
   (c) $M + N$. (A vector $\mathbf{v}$ belongs to $M + N$ if there are vectors $\mathbf{m} \in M$ and $\mathbf{n} \in N$ such that $\mathbf{v} = \mathbf{m} + \mathbf{n}$.)
   (d) $M - N$. (A vector $\mathbf{v}$ belongs to $M - N$ if there are vectors $\mathbf{m} \in M$ and $\mathbf{n} \in N$ such that $\mathbf{v} = \mathbf{m} - \mathbf{n}$.)

   Which of (a)–(d) are subspaces of V?
   Answer: _____ .

(6) For a fixed interval $[a, b]$, which sets of functions in the list 5.1.3 are vector subspaces of which?
Answer:

____ $\preceq$ ____ $\preceq$ ____ $\preceq$ ____ $\preceq$ ____ $\preceq$ ____ $\preceq$ ____ $\preceq$ ____ .

(7) Let M be the plane $x + y + z = 0$ and N be the line $x = y = z$ in $\mathbb{R}^3$. The purpose of this exercise is to confirm that $\mathbb{R}^3 = M \oplus N$. This requires establishing three things: (i) M and N are subspaces of $\mathbb{R}^3$ (which is *very* easy and which we omit); (ii) $\mathbb{R}^3 = M + N$; and (iii) $M \cap N = \{\mathbf{0}\}$.

(a) To show that $\mathbb{R}^3 = M + N$ we need $\mathbb{R}^3 \subseteq M + N$ and $M + N \subseteq \mathbb{R}^3$. Since $M \subseteq \mathbb{R}^3$ and $N \subseteq \mathbb{R}^3$, it is clear that $M + N \subseteq \mathbb{R}^3$. So all that is required is to show that $\mathbb{R}^3 \subseteq M + N$. That is, given a vector $\mathbf{x} = (x_1, x_2, x_3)$ in $\mathbb{R}^3$ we must find vectors $\mathbf{m} = (m_1, m_2, m_3)$ in M and $\mathbf{n} = (n_1, n_2, n_3)$ in N such that $\mathbf{x} = \mathbf{m} + \mathbf{n}$. Find two such vectors.

Answer: $\mathbf{m} = \dfrac{1}{3}$ (_____ , _____ ,
_____)

and $\mathbf{n} = \dfrac{1}{3}$ (_____ , _____ ,
_____).

(b) The last thing to verify is that $M \cap N = \{\mathbf{0}\}$; that is, that the only vector M and N have in common is the zero vector. Suppose that a vector $\mathbf{x} = (x_1, x_2, x_3)$ belongs to both M and N. Since $\mathbf{x} \in M$ it must satisfy the equation

$$x_1 + x_2 + x_3 = 0. \tag{1}$$

Since $x \in N$ it must satisfy the equations

$$x_1 = x_2 \quad \text{and} \tag{2}$$

$$x_2 = x_3. \tag{3}$$

Solve the system of equations (1)–(3).

Answer: $\mathbf{x} = ($ ____ , ____ , ____ $)$.

(8) Let $\mathcal{C} = \mathcal{C}[-1, 1]$ be the vector space of all continuous real valued functions on the interval $[-1, 1]$. A function f in $\mathcal{C}$ is EVEN if $f(-x) = f(x)$ for all $x \in [-1, 1]$; it is ODD if $f(-x) = -f(x)$ for all $x \in [-1, 1]$.

Let $C_o = \{f \in C : f \text{ is odd}\}$ and $C_e = \{f \in C : f \text{ is even}\}$. To show that $C = C_o \oplus C_e$ we need to show 3 things.

(i) C_o and C_e are subspaces of C. This is quite simple: let's do just one part of the proof. We will show that C_o is closed under addition. After each step of the following proof indicate the justification for that step. Make your choices from the following list.

- (A)   Arithmetic of real numbers.
- (DA)   Definition of addition of functions.
- (DE)   Definition of "even function".
- (DO)   Definition of "odd function".
- (H)   Hypothesis (that is, our assumptions or suppositions).

Proof. Let $f, g \in C_o$. Then

$$
\begin{aligned}
(f+g)(-x) &= f(-x) + g(-x) \quad &&(\text{reason: } \underline{\quad}) \\
&= -f(x) + (-g(x)) \quad &&(\text{reason: } \underline{\quad} \text{ and } \underline{\quad}) \\
&= -(f(x) + g(x)) \quad &&(\text{reason: } \underline{\quad}) \\
&= -(f+g)(x). \quad &&(\text{reason: } \underline{\quad})
\end{aligned}
$$

Thus $f + g \in C_o$. (reason $\underline{\quad}$).   □

(ii) $C_o \cap C_e = \{\mathbf{0}\}$ (where $\mathbf{0}$ is the constant function on $[-1, 1]$ whose value is zero). Again choose from the reasons listed in part (i) to justify the given proof.

Proof. Suppose $f \in C_o \cap C_e$. Then for each $x \in [-1, 1]$

$$
\begin{aligned}
f(x) &= f(-x) \quad &&(\text{reason: } \underline{\quad}) \\
&= -f(x). \quad &&(\text{reason: } \underline{\quad})
\end{aligned}
$$

Thus $f(x) = 0$ for every $x \in [-1, 1]$; that is, $f = \mathbf{0}$. (reason: $\underline{\quad}$).   □

(iii) $C = C_o + C_e$. To verify this we must show that every continuous function f on $[-1, 1]$ can be written as the sum of an odd function j and an even function k. It turns out that the functions j and k can be written as linear combinations of the given function f and the function g defined by $g(x) = f(-x)$ for $-1 \leq x \leq 1$. What are the appropriate coefficients?

$$\text{Answer:} \quad j = \underline{\quad} f + \underline{\quad} g$$
$$k = \underline{\quad} f + \underline{\quad} g.$$

(9) Let M be the line $x = y = z$ and N be the line $x = \frac{1}{2}y = \frac{1}{3}z$ in $\mathbb{R}^3$.

(a) The line M is the set of all scalar multiples of the vector $(1, \underline{\quad}, \underline{\quad})$.

(b) The line N is the set of all scalar multiples of the vector $(1, \underline{\quad}, \underline{\quad})$.

(c) The set $M + N$ is (geometrically speaking) a $\underline{\qquad}$ in $\mathbb{R}^3$; its equation is $ax + by + z = 0$ where $a = \underline{\quad}$ and $b = \underline{\quad}$.

(10) Let M be the plane $x - y + z = 0$ and N be the plane $x + 2y - z = 0$ in $\mathbb{R}^3$. State in one short sentence how you know that $\mathbb{R}^3$ is *not* the direct sum of M and N.

Answer: $\underline{\hspace{10cm}}$.

(11) Let M be the plane $2x - 3y + 4z + 1 = 0$ and N be the line $\frac{x}{4} = \frac{y}{2} = \frac{z}{3}$ in $\mathbb{R}^3$. State in one short sentence how you know that $\mathbb{R}^3$ is *not* the direct sum of M and N.

Answer: $\underline{\hspace{10cm}}$.

(12) Let M be the plane $x + y + z = 0$ and N be the line $x - 1 = \frac{1}{2}y = z + 2$ in $\mathbb{R}^3$. State in one short sentence how you know that $\mathbb{R}^3$ is *not* the direct sum of M and N.

Answer: $\underline{\hspace{10cm}}$.

(13) Let M be the line $x = y = z$ and N be the line $\frac{x}{4} = \frac{y}{2} = \frac{z}{3}$ in $\mathbb{R}^3$. State in one short sentence how you know that $\mathbb{R}^3$ is *not* the direct sum of M and N.

Answer: $\underline{\hspace{10cm}}$.

(14) Let M be the plane $x + y + z = 0$ and N be the line $x = -\frac{3}{4}y = 3z$. The purpose of this exercise is to see (in two different ways) that $\mathbb{R}^3$ is *not* the direct sum of M and N.

(a) If $\mathbb{R}^3$ were equal to $M \oplus N$, then $M \cap N$ would contain only the zero vector. Show that this is not the case by finding a nonzero vector $\mathbf{x}$ in $\mathbb{R}^3$ that belongs to $M \cap N$.

Answer: $\mathbf{x} = ($_____ , _____, $1)$.

(b) If $\mathbb{R}^3$ were equal to $M \oplus N$, then, in particular, we would have $\mathbb{R}^3 = M + N$. Since both M and N are subsets of $\mathbb{R}^3$, it is clear that $M + N \subseteq \mathbb{R}^3$. Show that the reverse inclusion $\mathbb{R}^3 \subseteq M + N$ is *not* correct by finding a vector $\mathbf{x} \in \mathbb{R}^3$ that cannot be written in the form $\mathbf{m} + \mathbf{n}$ where $\mathbf{m} \in M$ and $\mathbf{n} \in N$.

Answer: $\mathbf{x} = (-6, 8, a)$ is such a vector provided that $a \neq$ _____.

(c) We have seen in part (b) that $M + N \neq \mathbb{R}^3$. Then what *is* $M + N$?

Answer: $M + N =$ _____.

5.3 Problems

(1) Let M and N be subspaces of a vector space V. Consider the following subsets of V.

(a) $M \cap N$. (A vector $\mathbf{v}$ belongs to $M \cap N$ if it belongs to both M and N.)

(b) $M \cup N$. (A vector $\mathbf{v}$ belongs to $M \cup N$ if it belongs to either M or N.)

(c) $M + N$. (A vector $\mathbf{v}$ belongs to $M + N$ if there are vectors $\mathbf{m} \in M$ and $\mathbf{n} \in N$ such that $\mathbf{v} = \mathbf{m} + \mathbf{n}$.)

(d) $M - N$. (A vector $\mathbf{v}$ belongs to $M - N$ if there are vectors $\mathbf{m} \in M$ and $\mathbf{n} \in N$ such that $\mathbf{v} = \mathbf{m} - \mathbf{n}$.)

For each of the sets (a)–(d) above, either prove that it *is* a subspace of V or give a counterexample to show that it *need not* be a subspace of V.

(2) Let $\mathcal{C} = \mathcal{C}[0, 1]$ be the family of continuous real valued functions on the interval $[0, 1]$. Define

$$f_1(t) = t \quad \text{and} \quad f_2(t) = t^4$$

for $0 \le t \le 1$. Let M be the set of all functions of the form $\alpha f_1 + \beta f_2$ where $\alpha, \beta \in \mathbb{R}$. And let N be the set of all functions g in $\mathcal{C}$ that satisfy

$$\int_0^1 tg(t)\, dt = 0 \quad \text{and} \quad \int_0^1 t^4 g(t)\, dt = 0.$$

Is $\mathcal{C}$ the direct sum of M and N? (Give a careful proof of your claim and illustrate it with an example. What does your result say, for instance, about the function h defined by $h(t) = t^2$ for $0 \le t \le 1$.)

(3) Let V be a vector space.

    (a) Let $\mathcal{M}$ be a family of subspaces of V. Prove that the intersection $\bigcap \mathcal{M}$ of this family is itself a subspace of V.

    (b) Let A be a set of vectors in V. Explain carefully why it makes sense to say that the intersection of the family of all subspaces containing A is "the smallest subspace of V that contains A".

    (c) Prove that, for A as in part (b), the set of all sums of scalar multiples of vectors in A is the smallest subspace of V that contains A.

(4) In $\mathbb{R}^3$ let M be the line $x = y = z$, N be the line $x = \frac{1}{2}y = \frac{1}{3}z$, and $L = M + N$. Give a careful proof that $L = M \oplus N$.

(5) Let V be a vector space and suppose that $V = M \oplus N$. Show that for every $\mathbf{v} \in V$ there exist unique vectors $\mathbf{m} \in M$ and $\mathbf{n} \in N$ such that $\mathbf{v} = \mathbf{m} + \mathbf{n}$. *Hint.* It should be clear that the only thing you have to establish is the *uniqueness* of the vectors $\mathbf{m}$ and $\mathbf{n}$. To this end, suppose that a vector $\mathbf{v}$ in V can be written as $\mathbf{m}_1 + \mathbf{n}_1$ and it can also be written as $\mathbf{m}_2 + \mathbf{n}_2$ where $\mathbf{m}_1, \mathbf{m}_2 \in M$ and $\mathbf{n}_1, \mathbf{n}_2 \in N$. Prove that $\mathbf{m}_1 = \mathbf{m}_2$ and $\mathbf{n}_1 = \mathbf{n}_2$.

(6) Verify the assertions made in definition 5.1.7. In particular, show that $\sim$ is an equivalence relation, that addition and scalar multiplication of the set of equivalence classes are well defined, and that under these operations V/M is a vector space.

5.4   Answers to Odd-Numbered Exercises

(1) (d)

(3) $0, -2$

(5) (a), (c), and (d)

(7) (a) $2x_1 - x_2 - x_3$, $-x_1 + 2x_2 - x_3$, $-x_1 - x_2 + 2x_3$, $x_1 + x_2 + x_3$,
$x_1 + x_2 + x_3$, $x_1 + x_2 + x_3$,

(b) 0, 0, 0

(9) (a) 1, 1

(b) 2, 3

(c) plane, 1, -2

(11) M is not a subspace of $\mathbb{R}^3$.

(13) $M + N$ is a plane, not all of $\mathbb{R}^3$.

Chapter 6

LINEAR INDEPENDENCE

6.1 Background

Topics: linear combinations, span, linear dependence and independence.

Remark 6.1.1. Some authors of linear algebra texts make it appear as if the terms *linear dependence* and *linear independence*, *span*, and *basis* pertain only to *finite* sets of vectors. This is extremely misleading. The expressions should make sense for *arbitrary* sets. In particular, do not be misled into believing that a basis for a vector space must be a finite set of vectors (or a sequence of vectors). While it is true that in most elementary linear algebra courses the emphasis is on the study of *finite dimensional* vector spaces, bases for vector spaces may be *very* large indeed. I recommend the following definitions.

Definition 6.1.2. Recall that a vector $\mathbf{y}$ is a LINEAR COMBINATION of distinct vectors $\mathbf{x}_1, \ldots, \mathbf{x}_n$ if there exist scalars $\alpha_1, \ldots, \alpha_n$ such that $\mathbf{y} = \sum_{k=1}^{n} \alpha_k \mathbf{x}_k$. *Note:* linear combinations *are* finite sums. The linear combination $\sum_{k=1}^{n} \alpha_k \mathbf{x}_k$ is TRIVIAL if all the coefficients $\alpha_1, \ldots, \alpha_n$ are zero. If at least one α_k is different from zero, the linear combination is NONTRIVIAL.

Example 6.1.3. In $\mathbb{R}^2$ the vector $(8, 2)$ is a linear combination of the vectors $(1, 1)$ and $(1, -1)$ because $(8, 2) = 5(1, 1) + 3(1, -1)$.

Example 6.1.4. In $\mathbb{R}^3$ the vector $(1, 2, 3)$ is *not* a linear combination of the vectors $(1, 1, 0)$ and $(1, -1, 0)$.

Definition 6.1.5. Suppose that A is a subset (finite or not) of a vector space V. The SPAN of A is the set of all linear combinations of elements of A.

Another way of saying the same thing: the SPAN of A is the smallest subspace of V which contains A. (That these characterizations are equivalent is not completely obvious. Proof is required. See Problem 3 in Chapter 5.) We denote the span of A by span A. If $U = $ span A, we say that A SPANS U or that U is SPANNED BY A.

Example 6.1.6. For each $n = 0, 1, 2, \ldots$ define a function $\mathbf{p}_n$ on $\mathbb{R}$ by $\mathbf{p}_n(x) = x^n$. Let $\mathcal{P}$ be the set of polynomial functions on $\mathbb{R}$. It is a subspace of the vector space of continuous functions on $\mathbb{R}$. Then $\mathcal{P} = $ span$\{\mathbf{p}_0, \mathbf{p}_1, \mathbf{p}_2 \ldots\}$. The exponential function exp, whose value at x is e^x, is not in the span of the set $\{\mathbf{p}_0, \mathbf{p}_1, \mathbf{p}_2 \ldots\}$.

Definition 6.1.7. A subset A (finite or not) of a vector space is LINEARLY DEPENDENT if the zero vector $\mathbf{0}$ can be written as a nontrivial linear combination of elements of A; that is, if there exist distinct vectors $\mathbf{x}_1, \ldots, \mathbf{x}_n \in A$ and scalars $\alpha_1, \ldots, \alpha_n$, **not all zero**, such that $\sum_{k=1}^{n} \alpha_k \mathbf{x}_k = \mathbf{0}$. A subset of a vector space is LINEARLY INDEPENDENT if it is not linearly dependent.

Technically, it is a *set* of vectors that is linearly dependent or independent. Nevertheless, these terms are frequently used as if they were properties of the vectors themselves. For instance, if $S = \{\mathbf{x}_1, \ldots, \mathbf{x}_n\}$ is a finite set of vectors in a vector space, you may see the assertions "the set S is linearly independent" and "the vectors $\mathbf{x}_1, \ldots, \mathbf{x}_n$ are linearly independent" used interchangeably.

Example 6.1.8. The (vectors going from the origin to) points on the unit circle in $\mathbb{R}^2$ are linearly dependent. Reason: If $\mathbf{x} = (1, 0)$, $\mathbf{y} = \left(-\frac{1}{2}, \frac{\sqrt{3}}{2}\right)$, and $\mathbf{z} = \left(\frac{1}{2}, \frac{\sqrt{3}}{2}\right)$, then $\mathbf{x} + \mathbf{y} + (-1)\mathbf{z} = \mathbf{0}$.

Example 6.1.9. For each $n = 0, 1, 2, \ldots$ define a function $\mathbf{p}_n$ on $\mathbb{R}$ by $\mathbf{p}_n(x) = x^n$. Then the set $\{\mathbf{p}_0, \mathbf{p}_1, \mathbf{p}_2, \ldots\}$ is a linearly independent subset of the vector space of continuous functions on $\mathbb{R}$.

6.2 Exercises

(1) Show that in the space $\mathbb{R}^3$ the vectors $\mathbf{x} = (1, 1, 0)$, $\mathbf{y} = (0, 1, 2)$, and $\mathbf{z} = (3, 1, -4)$ are linearly dependent by finding scalars α and β such that $\alpha \mathbf{x} + \beta \mathbf{y} + \mathbf{z} = \mathbf{0}$.

Answer: $\alpha = $ _____, $\beta = $ _____.

(2) Let $\mathbf{w} = (1, 1, 0, 0)$, $\mathbf{x} = (1, 0, 1, 0)$, $\mathbf{y} = (0, 0, 1, 1)$, and $\mathbf{z} = (0, 1, 0, 1)$.

(a) We can show that $\{\mathbf{w}, \mathbf{x}, \mathbf{y}, \mathbf{z}\}$ is not a spanning set for $\mathbb{R}^4$ by finding a vector $\mathbf{u}$ in $\mathbb{R}^4$ such that $\mathbf{u} \notin \text{span}\{\mathbf{w}, \mathbf{x}, \mathbf{y}, \mathbf{z}\}$. One such vector is $\mathbf{u} = (1, 2, 3, a)$ where a is any number *except* _____.

(b) Show that $\{\mathbf{w}, \mathbf{x}, \mathbf{y}, \mathbf{z}\}$ is a linearly dependent set of vectors by finding scalars α, γ, and δ such that $\alpha\mathbf{w} + \mathbf{x} + \gamma\mathbf{y} + \delta\mathbf{z} = \mathbf{0}$.
Answer: $\alpha = $ _____, $\gamma = $ _____, $\delta = $ _____.

(c) Show that $\{\mathbf{w}, \mathbf{x}, \mathbf{y}, \mathbf{z}\}$ is a linearly dependent set by writing $\mathbf{z}$ as a linear combination of $\mathbf{w}$, $\mathbf{x}$, and $\mathbf{y}$. Answer: $\mathbf{z} = $ ___ $\mathbf{w} + $ _____ $\mathbf{x} + $ ___ $\mathbf{y}$.

(3) Let $p(x) = x^2 + 2x - 3$, $q(x) = 2x^2 - 3x + 4$, and $r(x) = ax^2 - 1$. The set $\{p, q, r\}$ is linearly dependent if $a = $ _____.

(4) Show that in the vector space $\mathbb{R}^3$ the vectors $\mathbf{x} = (1, 2, -1)$, $\mathbf{y} = (3, 1, 1)$, and $\mathbf{z} = (5, -5, 7)$ are linearly dependent by finding scalars α and β such that $\alpha\mathbf{x} + \beta\mathbf{y} + \mathbf{z} = \mathbf{0}$.

Answer: $\alpha = $ _____, $\beta = $ _____.

(5) Let $f_1(x) = \sin x$, $f_2(x) = \cos(x + \pi/6)$, and $f_3(x) = \sin(x - \pi/4)$ for $0 \leq x \leq 2\pi$. Show that $\{f_1, f_2, f_3\}$ is linearly dependent by finding constants α and β such that $\alpha f_1 - 2f_2 - \beta f_3 = \mathbf{0}$.

Answer: $\alpha = $ _____ and $\beta = $ _____.

(6) In the space $\mathcal{C}[0, \pi]$ let f, g, h, and j be the vectors defined by

$$f(x) = 1$$
$$g(x) = x$$
$$h(x) = \cos x$$
$$j(x) = \cos^2 \frac{x}{2}$$

for $0 \leq x \leq \pi$. Show that f, g, h, and j are linearly dependent by writing j as a linear combination of f, g, and h.
Answer: $j = $ _____ $f + $ _____ $g + $ _____ h.

(7) Let $\mathbf{u} = (\lambda, 1, 0)$, $\mathbf{v} = (1, \lambda, 1)$, and $\mathbf{w} = (0, 1, \lambda)$. Find **all** values of λ which make $\{\mathbf{u}, \mathbf{v}, \mathbf{w}\}$ a linearly dependent subset of $\mathbb{R}^3$. Answer:

_____.

(8) Let $\mathbf{u} = (1, 0, -2)$, $\mathbf{v} = (1, 2, \lambda)$, and $\mathbf{w} = (2, 1, -1)$. Find **all** values of λ which make $\{\mathbf{u}, \mathbf{v}, \mathbf{w}\}$ a linearly dependent subset of $\mathbb{R}^3$. Answer:

_____.

(9) Let $p(x) = x^3 - x^2 + 2x + 3$, $q(x) = 3x^3 + x^2 - x - 1$, $r(x) = x^3 + 2x + 2$, and $s(x) = 7x^3 + ax^2 + 5$. The set $\{p, q, r, s\}$ is linearly dependent if $a = \underline{\quad}$.

(10) In the space $\mathcal{C}[0, \pi]$ define the vectors f, g, and h by

$$f(x) = x$$
$$g(x) = \sin x$$
$$h(x) = \cos x$$

for $0 \leq x \leq \pi$. We show that f, g, and h are linearly independent. This is accomplished by showing that if $\alpha f + \beta g + \gamma h = \mathbf{0}$, then $\alpha = \beta = \gamma = 0$. So we start by supposing that $\alpha f + \beta g + \gamma h = \mathbf{0}$; that is,

$$\alpha x + \beta \sin x + \gamma \cos x = 0 \tag{1}$$

for all $x \in [0, \pi]$.

(a) We see that γ must be zero by setting $x = \underline{\quad}$ in equation (1).

Now differentiate (1) to obtain

$$\alpha + \beta \cos x = 0 \tag{2}$$

for all $x \in [0, \pi]$.

(b) We see that α must be zero by setting $x = \underline{\quad}$ in equation (2).

Differentiate (2) to obtain

$$-\beta \sin x = 0 \tag{3}$$

for all $x \in [0, \pi]$.

(c) We conclude that $\beta = 0$ by setting $x = \underline{\quad}$ in (3).

6.3 Problems

(1) In the space $C[0, 1]$ define the vectors f, g, and h by

$$f(x) = x$$
$$g(x) = e^x$$
$$h(x) = e^{-x}$$

for $0 \leq x \leq 1$. Use the **definition** of *linear independence* to show that the functions f, g, and h are linearly independent.

(2) Let a, b, and c be distinct real numbers. Use the **definition** of *linear independence* to give a careful proof that the vectors $(1, 1, 1)$, (a, b, c), and (a^2, b^2, c^2) form a linearly independent subset of $\mathbb{R}^3$.

(3) Let $\{\mathbf{u}, \mathbf{v}, \mathbf{w}\}$ be a linearly independent set in a vector space V. Use the **definition** of *linear independence* to give a careful proof that the set $\{\mathbf{u} + \mathbf{v}, \mathbf{u} + \mathbf{w}, \mathbf{v} + \mathbf{w}\}$ is linearly independent in V.

(4) You are the leader of an engineering group in the company you work for and have a routine computation that has to be done repeatedly. At your disposal is an intern, Kim, a beginning high school student, who is bright but has had no advanced mathematics. In particular, Kim knows nothing about vectors or matrices.

Here is the computation that is needed. Three vectors, $\mathbf{a}$, $\mathbf{b}$, and $\mathbf{c}$ are specified in $\mathbb{R}^5$. (Denote their span by M.) Also specified is a (sometimes long) list of other vectors $S = \{\mathbf{v}_1, \mathbf{v}_2, \ldots, \mathbf{v}_n\}$ in $\mathbb{R}^5$. The problem is to

(1) determine which of the vectors in S belong to M, and
(2) for each vector $\mathbf{v}_k \in S$ which *does* belong to M

find constants α, β, and γ such that $\mathbf{v}_k = \alpha\mathbf{a} + \beta\mathbf{b} + \gamma\mathbf{c}$.

Kim has access to Computer Algebra System (Maple, or a similar program) with a Linear Algebra package. Write a simple and efficient algorithm (that is, a set of instructions) which will allow Kim to carry out the desired computation repeatedly. The algorithm should be simple in the sense that it uses only the most basic linear algebra commands (for example, *Matrix, Vector, Transpose, RowReducedEchelonForm, etc.* in Maple). Remember, you must tell Kim everything: how to set up the appropriate matrices, what operations to perform on them, and how to interpret the results. The algorithm should be as efficient as you can make it. For example, it would certainly not be efficient for Kim to retype the coordinates of **a**, **b**, and **c** for each new $\mathbf{v}_k$.

Include in your write-up an actual printout showing how your algorithm works in some special case of your own invention. (For this example, the set S need contain only 5 or 6 vectors, some in U, some not.)

(5) The point of this problem is not just to get a correct answer to (a)–(c) below using tools you may have learned elsewhere, but *to give a careful explanation of how to apply the linear algebra techniques you have already encountered to solve this problem in a systematic fashion.* For background you may wish to read a bit about networks and *Kirchhoff's laws* (see, for example, [5] *Topic: Analyzing Networks*, pages 72–77 or [1] *Electrical Networks*, pages 538–542).

Consider an electrical network having four nodes **A**, **B**, **C**, and **D** connected by six branches **1**, ..., **6**. Branch **1** connects **A** and **B**; branch **2** connects **B** and **D**; branch **3** connects **C** and **B**; branch **4** connects **C** and **D**; branch **5** connects **A** and **C**; and branch **6** connects **A** and **D**.

The current in branch **k** is I_k, where $k = 1, \ldots, 6$. There is a 17 volt battery in branch **1** producing the current I_1 that flows from **A** to **B**. In branches **2**, **4**, and **5** there are 0.5 ohm resistors; and in branches **1**, **3**, and **6** there are 1 ohm resistors.

(a) Find the current in each branch. (Explain any minus signs that occur in your answer.)

(b) Find the voltage drop across each branch.

(c) Let $p_{\mathbf{n}}$ be the potential at node $\mathbf{n} = \mathbf{A}, \mathbf{B}, \mathbf{C}, \mathbf{D}$. The voltage drop across the branch connecting node **j** to node **k** is the difference in the potentials at nodes **j** and **k**. Suppose the network is grounded at **D** (so that $p_{\mathbf{D}} = 0$). Find the potential at the other nodes.

6.4   Answers to Odd-Numbered Exercises

(1) $-3, 2$

(3) 7

(5) $\sqrt{3} - 1, \sqrt{6}$

(7) $-\sqrt{2}, 0, \sqrt{2}$

(9) -3

Chapter 7

BASIS FOR A VECTOR SPACE

7.1 Background

Topics: basis, dimension.

Definition 7.1.1. A set B (finite or not) of vectors in a vector space V is a BASIS for V if it is linearly independent and spans V.

Example 7.1.2. The vectors $\mathbf{e}^1 = (1,0,0)$, $\mathbf{e}^2 = (0,1,0)$, and $\mathbf{e}^3 = (0,0,1)$ constitute a basis for the vector space $\mathbb{R}^3$.

Example 7.1.3. More generally, consider the vector space $\mathbb{R}^n$ of all n-tuples of real numbers. For each natural number k between 1 and n let $\mathbf{e}^k$ be the vector that is 1 in the kth-coordinate and 0 in all the others. Then the set $\{\mathbf{e}^1, \mathbf{e}^2, \ldots, \mathbf{e}^n\}$ is a basis for $\mathbb{R}^n$. It is called the STANDARD BASIS for $\mathbb{R}^n$.

Example 7.1.4. For each $n = 0,1,2,\ldots$ define a function $\mathbf{p}_n$ on $\mathbb{R}$ by $\mathbf{p}_n(x) = x^n$. Then the set $\{\mathbf{p}_0, \mathbf{p}_1, \mathbf{p}_2, \ldots\}$ is a basis for the vector space $\mathcal{P}$ of polynomial functions on $\mathbb{R}$.

Two important facts of linear algebra are that regardless of the size of the space *every* vector space has a basis and that every subspace has a complement.

Theorem 7.1.5. *Let B be a linearly independent set of vectors in a vector space V. Then there exists a set C of vectors in V such that $B \cup C$ is a basis for V.*

Corollary 7.1.6. *Every vector space has a basis.*

Corollary 7.1.7. *Let V be a vector space. If $M \preceq V$, then there exists $N \preceq V$ such that $M \oplus N = V$.*

The next theorem says that any two bases for a vector space are the same size.

Theorem 7.1.8. *If B and C are bases for the same vector space, then there is a one-to-one correspondence from B onto C.*

Definition 7.1.9. A vector space V is FINITE DIMENSIONAL if it has a finite basis. Its DIMENSION (denoted by $\dim V$) is the number of elements in the basis. If V does not have a finite basis it is INFINITE DIMENSIONAL.

Remark 7.1.10. Although most elementary texts contain proofs of Theorem 7.1.5 (or its corollaries) and Theorem 7.1.8 for finite dimensional spaces, it is difficult to find in them proofs that hold for infinite dimensional spaces. The reason for this is that such proofs require a set theoretic axiom called *Zorn's Lemma*, which many instructors feel is a topic not appropriate for beginning courses. (See the discussion in [7], p. 238, *ff.*)

Theorem 7.1.11. *If M and N are finite dimensional subspaces of a vector space, then $M + N$ is finite dimensional and*

$$\dim(M + N) = \dim M + \dim N - \dim(M \cap N).$$

7.2 Exercises

(1) Let $\mathbf{u} = (2, 0, -1)$, $\mathbf{v} = (3, 1, 0)$, and $\mathbf{w} = (1, -1, c)$ where $c \in \mathbb{R}$. The set $\{\mathbf{u}, \mathbf{v}, \mathbf{w}\}$ is a basis for $\mathbb{R}^3$ provided that c is *not* equal to _____.

(2) Let $\mathbf{u} = (1, -1, 3)$, $\mathbf{v} = (1, 0, 1)$, and $\mathbf{w} = (1, 2, c)$ where $c \in \mathbb{R}$. The set $\{\mathbf{u}, \mathbf{v}, \mathbf{w}\}$ is a basis for $\mathbb{R}^3$ provided that c is *not* equal to _____.

(3) The dimension of $\mathfrak{M}_{2 \times 2}$, the vector space of all 2×2 matrices of real numbers is _____.

(4) The dimension of $\mathfrak{T}_2$, the vector space of all 2×2 matrices of real numbers with zero trace is _____.

(5) The dimension of the vector space of all real valued polynomial functions on $\mathbb{R}$ of degree 4 or less is _____.

(6) In $\mathbb{R}^4$ let M be the subspace spanned by the vectors $(1, 1, 1, 0)$ and $(0, -4, 1, 5)$ and let N be the subspace spanned by $(0, -2, 1, 2)$ and

$(1, -1, 1, 3)$. One vector that belongs to both M and N is $(1, \underline{\hspace{1cm}}, \underline{\hspace{1cm}}, \underline{\hspace{1cm}})$. The dimension of $M \cap N$ is $\underline{\hspace{1cm}}$ and the dimension of $M + N$ is $\underline{\hspace{1cm}}$.

7.3 Problems

(1) Exhibit a basis for $\mathfrak{M}_{2 \times 2}$, the vector space of all 2×2 matrices of real numbers.

(2) Exhibit a basis for $\mathfrak{T}_2$, the vector space of all 2×2 matrices of real numbers with zero trace.

(3) Exhibit a basis for $\mathfrak{S}_3$, the vector space of all symmetric 3×3 matrices of real numbers.

(4) Let $\mathfrak{U}$ be the set of all matrices of real numbers of the form $\begin{bmatrix} u & -u - x \\ 0 & x \end{bmatrix}$ and $\mathfrak{V}$ be the set of all real matrices of the form $\begin{bmatrix} v & 0 \\ w & -v \end{bmatrix}$. Exhibit a basis for $\mathfrak{U}$, for $\mathfrak{V}$, for $\mathfrak{U} + \mathfrak{V}$, and for $\mathfrak{U} \cap \mathfrak{V}$.

(5) Prove that the vectors $(1, 1, 0)$, $(1, 2, 3)$, and $(2, -1, 5)$ form a basis for $\mathbb{R}^3$.

(6) Let V be a vector space and A be a linearly independent subset of V. Prove that A is a basis for V if and only if it is a maximal linearly independent subset of V. (If A is a linearly independent subset of V we say that it is a MAXIMAL linearly independent set if the addition of any vector at all to A will result in a set that is *not* linearly independent.)

(7) Let V be a vector space and A a subset of V that spans V. Prove that A is a basis for V if and only if it is a minimal spanning set. (If A is a set which spans V we say that it is a MINIMAL spanning set if the removal of any vector at all from A will result in a set which does *not* span V.)

7.4 Answers to Odd-Numbered Exercises

(1) -2

(3) 4

(5) 5

Part 3

LINEAR MAPS BETWEEN VECTOR SPACES

Chapter 8

LINEARITY

8.1 Background

Topics: linear maps between vector spaces, kernel, nullspace, nullity, range, rank, isomorphism.

Definition 8.1.1. A function $f\colon A \to B$ is ONE-TO-ONE (or INJECTIVE) if $u = v$ in A whenever $f(u) = f(v)$ in B.

Definition 8.1.2. A function $f\colon A \to B$ is ONTO (or SURJECTIVE) if for every $b \in B$ there exists $a \in A$ such that $b = f(a)$.

Definition 8.1.3. A function $f\colon A \to B$ is a ONE-TO-ONE CORRESPONDENCE (or BIJECTIVE) if it is both injective and surjective (one-to-one and onto).

Definition 8.1.4. A map $T\colon V \to W$ between vector spaces is LINEAR if

$$T(\mathbf{x} + \mathbf{y}) = T\mathbf{x} + T\mathbf{y} \quad \text{for all } \mathbf{x}, \mathbf{y} \in V \tag{8.1.1}$$

and

$$T(\alpha\mathbf{x}) = \alpha T\mathbf{x} \quad \text{for all } \mathbf{x} \in V \text{ and } \alpha \in \mathbb{F}. \tag{8.1.2}$$

Here $\mathbb{F} = \mathbb{R}$ if V and W are real vector spaces and $\mathbb{F} = \mathbb{C}$ if they are complex vector spaces.

A scalar valued linear map on a vector space V is a LINEAR FUNCTIONAL.

A linear map is frequently called a LINEAR TRANSFORMATION, and, in case the domain and codomain are the same, it is often called a (LINEAR) OPERATOR. The family of all linear transformations from V into W is denoted by $\mathfrak{L}(V, W)$. We shorten $\mathfrak{L}(V, V)$ to $\mathfrak{L}(V)$.

Two oddities of notation concerning linear transformations deserve comment. First, the value of T at $\mathbf{x}$ is usually written $T\mathbf{x}$ rather than $T(\mathbf{x})$. Naturally the parentheses are used whenever their omission would create ambiguity. For example, in (8.1.1) above $T\mathbf{x} + \mathbf{y}$ is not an acceptable substitute for $T(\mathbf{x} + \mathbf{y})$. Second, the symbol for composition of two linear transformations is ordinarily omitted. If $S \in \mathfrak{L}(U, V)$ and $T \in \mathfrak{L}(V, W)$, then the composite of T and S is denoted by TS (rather than by $T \circ S$). As a consequence of this convention when $T \in \mathfrak{L}(V)$ the linear operator $T \circ T$ is written as T^2, $T \circ T \circ T$ as T^3, and so on.

For future reference here are two obvious properties of a linear map.

Proposition 8.1.5. *If $T : V \to W$ is a linear map between vector spaces, then $T(\mathbf{0}) = \mathbf{0}$.*

Proposition 8.1.6. *If $T : V \to W$ is a linear map between vector spaces, then $T(-\mathbf{x}) = -T\mathbf{x}$ for every $\mathbf{x} \in V$.*

You should prove these propositions if (and only if) it is not immediately obvious to you how to do so.

Definition 8.1.7. Let $T : V \to W$ be a linear transformation between vector spaces. Then $\ker T$, the KERNEL of T, is defined to be the set of all $\mathbf{x}$ in V such that $T\mathbf{x} = \mathbf{0}$. The kernel of T is also called the *nullspace* of T. If V is finite dimensional, the dimension of the kernel of T is the NULLITY of T.

Also, $\operatorname{ran} T$, the RANGE of T, is the set of all $\mathbf{y}$ in W such that $\mathbf{y} = T\mathbf{x}$ for some $\mathbf{x}$ in V. If the range of T is finite dimensional, its dimension is the RANK of T.

Notation 8.1.8. Let V be a vector space. We denote the IDENTITY MAP on V (that is, the map $\mathbf{x} \mapsto \mathbf{x}$ from V into itself) by I_V, or just I.

The following fundamental result is proved in most linear algebra texts.

Theorem 8.1.9. *If $T : V \to W$ is a linear map between finite dimensional vector spaces, then*

$$\operatorname{rank}(T) + \operatorname{nullity}(T) = \dim V.$$

(For an example of a proof, see [4], p. 90.)

Definition 8.1.10. Let $T\colon V \to W$ and $S\colon W \to V$ be linear maps. If $ST = I_V$, then T is a RIGHT INVERSE for S and S is a LEFT INVERSE for T. The mapping T is INVERTIBLE (or is an ISOMORPHISM) if there exists a linear transformation, which we denote by T^{-1} mapping W into V, such that

$$TT^{-1} = I_W \quad \text{and} \quad T^{-1}T = I_V.$$

The vector spaces V and W are ISOMORPHIC if there exists an isomorphism T from V to W. We write $V \cong W$ to indicate that V and W are isomorphic.

8.2 Exercises

(1) Define $T\colon \mathbb{R}^3 \to \mathbb{R}^4$ by

$$T\mathbf{x} = (x_1 - x_3, x_1 + x_2, x_3 - x_2, x_1 - 2x_2)$$

for all $\mathbf{x} = (x_1, x_2, x_3)$ in $\mathbb{R}^3$.

(a) Then $T(1, -2, 3) = ($ ____, ____, ____, ____ $)$.

(b) Find a vector $\mathbf{x} \in \mathbb{R}^3$ such that $T\mathbf{x} = (8, 9, -5, 0)$.

Answer: $\mathbf{x} = ($ ____, ____, ____ $)$.

(2) Define $T\colon \mathbb{R}^4 \to \mathbb{R}^3$ by

$$T\mathbf{x} = (2x_1 + x_3 + x_4, x_1 - 2x_2 - x_3, x_2 - x_3 + x_4)$$

for all $\mathbf{x} = (x_1, x_2, x_3, x_4)$ in $\mathbb{R}^4$.

(a) Then $T(2, 1, -1, 3) = ($ ____, ____, ____ $)$.

(b) Find a vector $\mathbf{x} \in \mathbb{R}^4$ such that $T\mathbf{x} = (3, -1, -3)$.

Answer: $\mathbf{x} = ($ ____, ____, ____, ____ $)$.

(3) Let T be the linear map from $\mathbb{R}^3$ to $\mathbb{R}^3$ defined by

$$T(x, y, z) = (x + 2y - z, 2x + 3y + z, 4x + 7y - z).$$

The kernel of T is (geometrically) a _____ whose equation(s) is(are)

_____ ;

and the range of T is geometrically a _____ whose equation(s) is(are) _____ .

(4) Let $T\colon \mathbb{R}^3 \to \mathbb{R}^3$ be the linear transformation whose action on the standard basis vectors of $\mathbb{R}^3$ is

$$T(1,0,0) = \left(1, -\frac{3}{2}, 2\right)$$

$$T(0,1,0) = \left(-3, \frac{9}{2}, -6\right)$$

$$T(0,0,1) = (2, -3, 4).$$

Then $T(5,1,-1) = (\underline{\quad}, \underline{\quad}, \underline{\quad})$. The kernel of T is the _____ whose equation is $x + ay + bz = 0$ where $a = \underline{\quad}$ and $b = \underline{\quad}$. The range of T is the _____ whose equations are $\dfrac{x}{2} = \dfrac{y}{c} = \dfrac{z}{d}$ where $c = \underline{\quad}$ and where $d = \underline{\quad}$.

(5) Let $\mathcal{P}$ be the vector space of all polynomial functions on $\mathbb{R}$ with real coefficients. Define linear transformations $T,\ D\colon \mathcal{P} \to \mathcal{P}$ by

$$(Dp)(x) = p'(x)$$

and

$$(Tp)(x) = xp(x)$$

for all $x \in \mathbb{R}$.

(a) Let $p(x) = x^3 - 7x^2 + 5x + 6$ for all $x \in \mathbb{R}$. Then $((D+T)(p))(x) = x^4 - ax^3 + bx^2 - bx + c$ where $a = \underline{\quad}$, $b = \underline{\quad}$, and $c = \underline{\quad}$.
(b) Let p be as in (a). Then $(DTp)(x) = ax^3 - bx^2 + cx + 6$ where $a = \underline{\quad}$, $b = \underline{\quad}$, and $c = \underline{\quad}$.
(c) Let p be as in (a). Then $(TDp)(x) = ax^3 - bx^2 + cx$ where $a = \underline{\quad}$, $b = \underline{\quad}$, and $c = \underline{\quad}$.
(d) Evaluate (and simplify) the commutator $[D,T] := DT - TD$. Answer: $[D,T] = \underline{\quad\quad}$.
(e) Find a number p such that $(TD)^p = T^p D^p + TD$. Answer: $p = \underline{\quad\quad}$.

(6) Let $\mathcal{C} = \mathcal{C}[a,b]$ be the vector space of all continuous real valued functions on the interval $[a,b]$ and $\mathcal{C}^1 = \mathcal{C}^1[a,b]$ be the vector space of all continuously differentiable real valued functions on $[a,b]$. (Recall that a function is CONTINUOUSLY DIFFERENTIABLE if it has a derivative

and the derivative is continuous.) Let $D: C^1 \to C$ be the linear transformation defined by

$$Df = f'$$

and let $T: C \to C^1$ be the linear transformation defined by

$$(Tf)(x) = \int_a^x f(t)\, dt$$

for all $f \in C$ and $x \in [a, b]$.

(a) Compute (and simplify) $(DTf)(x)$. Answer: _____.

(b) Compute (and simplify) $(TDf)(x)$. Answer: _____.

(c) The kernel of T is _____.

(d) The range of T is $\{g \in C^1: _____\}$.

(7) In this exercise we prove that a linear transformation $T: V \to W$ between two vector spaces is one-to-one if and only if its kernel contains only the zero vector. After each step of the proof give the appropriate reason. Choose reasons from the following list.

(DK) Definition of "kernel".

(DL) Definition of "linear".

(DO) Definition of "one-to-one".

 (H) Hypothesis.

(Pa) Proposition 8.1.5.

(Pb) Proposition 8.1.6.

(VA) Vector space arithmetic (consequences of vector space axioms, definition of subtraction of vectors, etc.)

Proof. Suppose that T is one-to-one. We show that $\ker T = \{\mathbf{0}_V\}$. Since $\mathbf{0}_V \in \ker T$ (reason: _____ and _____), we need only show that $\ker T \subseteq \{\mathbf{0}_V\}$; that is, we show that if $\mathbf{x} \in \ker T$, then $\mathbf{x} = \mathbf{0}_V$. So let $\mathbf{x} \in \ker T$. Then $T\mathbf{x} = \mathbf{0}_W$ (reason: _____ and _____) and $T\mathbf{0}_V = \mathbf{0}_W$ (reason: _____). From this we conclude that $\mathbf{x} = \mathbf{0}_V$ (reason: _____ and _____).

Now we prove the converse. Suppose that $\ker T = \{\mathbf{0}_V\}$. We wish to show that T is one-to-one. Let $\mathbf{x}, \mathbf{y} \in V$ and suppose that $T\mathbf{x} = T\mathbf{y}$.

Then

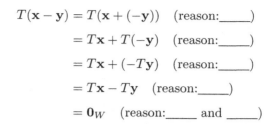

$$T(\mathbf{x} - \mathbf{y}) = T(\mathbf{x} + (-\mathbf{y})) \quad (\text{reason:}\_\_\_\_)$$
$$= T\mathbf{x} + T(-\mathbf{y}) \quad (\text{reason:}\_\_\_\_)$$
$$= T\mathbf{x} + (-T\mathbf{y}) \quad (\text{reason:}\_\_\_\_)$$
$$= T\mathbf{x} - T\mathbf{y} \quad (\text{reason:}\_\_\_\_)$$
$$= \mathbf{0}_W \quad (\text{reason:}\_\_\_\_ \text{ and } \_\_\_\_)$$

Then $\mathbf{x} - \mathbf{y} \in \ker T$ (reason: ____). So $\mathbf{x} - \mathbf{y} = \mathbf{0}_V$ (reason: ____); that is, $\mathbf{x} = \mathbf{y}$ (reason: ____). Thus T is one-to-one (reason: ____ and ____). □

(8) Let $\mathcal{C}^1(\mathbb{R})$ be the vector space of all functions defined on the real line $\mathbb{R}$ that have continuous derivatives at each point of $\mathbb{R}$ and $\mathcal{C}(\mathbb{R})$ be the vector space of continuous functions on $\mathbb{R}$. Define the function $T: \mathcal{C}^1(\mathbb{R}) \to \mathcal{C}(\mathbb{R})$ by

$$(Tf)(t) = f'(t) + 3f(t)$$

for every $t \in \mathbb{R}$. (Notice that T is a linear map.) The kernel of T is the set of all scalar multiples of the function g where $g(t) = _____$ for each t. Thus the kernel of the linear map T is the solution space of the differential equation _____.

(9) Let $\mathcal{C}^2(\mathbb{R})$ be the vector space of all functions defined on the real line $\mathbb{R}$ that have continuous second derivatives at each point of $\mathbb{R}$ and $\mathcal{C}(\mathbb{R})$ be the vector space of continuous functions on $\mathbb{R}$. Define the function $T: \mathcal{C}^2(\mathbb{R}) \to \mathcal{C}(\mathbb{R})$ by

$$(Tf)(t) = f''(t) + f(t)$$

for every $t \in \mathbb{R}$. (Notice that T is a linear map.) Assume that the kernel of T is two dimensional. Then $\ker T = \operatorname{span}\{g, h\}$ where $g(t) = _____$ and $h(t) = _____$ for all t. Thus the kernel of the linear map T is the solution space of the differential equation _____.

(10) Define a function k on the unit square $[0,1] \times [0,1]$ by

$$k(x,y) = \begin{cases} x, & \text{for } 0 \le x \le y \le 1 \\ y, & \text{for } 0 \le y < x \le 1 \end{cases}.$$

Define an integral operator K on the vector space $\mathcal{C}[0,1]$ of continuous real valued functions on $[0,1]$ by

$$(Kf)(x) = \int_0^1 k(x,y)f(y)\,dy$$

for $0 \le x \le 1$. Find the function Kf when f is the function defined by $f(x) = x^2$ for $0 \le x \le 1$.

Answer: $(Kf)(x) = $ _____.

(11) Let $T\colon \mathbb{R}^3 \to \mathbb{R}^3 \colon \mathbf{x} \mapsto (x_1 + 3x_2 - 2x_3, x_1 - 4x_3, x_1 + 6x_2)$.

(a) The kernel of T is a _____ in $\mathbb{R}^3$ given by the equation(s) _____.

(b) The range of T is a _____ in $\mathbb{R}^3$ given by the equation(s) _____.

(12) Let $T\colon \mathbb{R}^2 \to \mathbb{R}^3 \colon (x,y) \mapsto (2x - 3y, x + 2y + 1, 5x - 2y)$. State in one short sentence how you know that T is *not* a linear transformation.

Answer: _____.

(13) Let $\mathbf{a} = (1,0,0,0)$, $\mathbf{b} = (1,1,0,0)$, $\mathbf{c} = (1,1,1,0)$, and $\mathbf{d} = (1,1,1,1)$. Suppose that $T\colon \mathbb{R}^4 \to \mathbb{R}^7$ is a mapping such that $T(\mathbf{a}) = T(\mathbf{b}) = T(\mathbf{c}) = T(\mathbf{d}) = \mathbf{0}$ and that $T(3, -19, 7, -8) = (1,1,1,-3,6,2,5)$. State in a short sentence or two how you know that T is *not* a linear transformation.

Answer: _____.

(14) Suppose that $T\colon \mathbb{R}^3 \to \mathbb{R}^3$ is a mapping (not identically zero) whose range is contained in the paraboloid $z = x^2 + y^2$. State in a short sentence or two how you know that T is *not* a linear transformation.

Answer: _____.

(15) Let $T: \mathbb{R}^2 \to \mathbb{R}^4: (x, y) \mapsto (2x-3y, x-7y, x+2y+1, 5x-2y)$. State in one short sentence how you know that T is *not* a linear transformation.

Answer: _____.

(16) Let $\mathbf{a} = (1, 1, 0)$ and $\mathbf{b} = (0, 1, 1)$, and $\mathbf{c} = (1, 2, 1)$. Suppose that $T: \mathbb{R}^3 \to \mathbb{R}^5$ is a mapping such that $T(\mathbf{a}) = T(\mathbf{b}) = \mathbf{0}$ and that $T(\mathbf{c}) = (1, -3, 6, 2, 5)$. State in a short sentence or two how you know that T is *not* a linear transformation.

Answer: _____.

(17) Suppose that $T: \mathbb{R}^2 \to \mathbb{R}^2$ is a mapping (not identically zero) such that $T(1, 1) = (3, -6)$ and $T(-2, -2) = (-6, 3)$. State in a short sentence or two how you know that T is *not* a linear transformation.

Answer: _____.

8.3 Problems

(1) Let $T: V \to W$ be a linear transformation between vector spaces and let N be a subspace of W. Define $T^{\leftarrow}(N) := \{\mathbf{v} \in V: T\mathbf{v} \in N\}$. Prove that $T^{\leftarrow}(N)$ is a subspace of V.

(2) Prove that a linear transformation $T: \mathbb{R}^3 \to \mathbb{R}^2$ cannot be one-to-one and that a linear transformation $S: \mathbb{R}^2 \to \mathbb{R}^3$ cannot be onto. Generalize these assertions.

(3) Prove that one-to-one linear transformations preserve linear independence. That is: Let $T: V \to W$ be a one-to-one linear transformation between vector spaces and $\{\mathbf{x}_1, \mathbf{x}_2, \ldots, \mathbf{x}_n\}$ be a linearly independent subset of V. Prove that $\{T\mathbf{x}_1, T\mathbf{x}_2, \ldots, T\mathbf{x}_n\}$ is a linearly independent subset of W. *Hint.* To prove that the vectors $T\mathbf{x}_1, T\mathbf{x}_2, \ldots, T\mathbf{x}_n$ are linearly independent, it must be shown that the only linear combination of these vectors which equals zero is the trivial linear combination. So *suppose* that $\sum_{k=1}^{n} \alpha_k T\mathbf{x}_k = \mathbf{0}$ and *prove* that every α_k must be zero. Use the result proved in exercise 7.

(4) The goal of this problem is to understand and write up an introduction to *invertible* linear transformations. Your write-up should explain with spectacular clarity the basic facts about invertible linear transformations. Include answers to the following questions — giving complete proofs or counterexamples as required. (But don't number things in your report to correspond with the items that follow.)

(a) If a linear transformation has a right inverse must it have a left inverse?

(b) If a linear transformation has a left inverse must it have a right inverse?

(c) If a linear transformation has both a left and a right inverse, must it be invertible? (That is, must the left and right inverse be the same?)

(d) If a linear transformation T has a unique right inverse is T necessarily invertible? *Hint.* Consider $ST + S - I$, where S is a unique right inverse for T.

(e) What is the relationship between a linear transformation being one-to-one and onto and being invertible?

(f) Let $\{\mathbf{v}_1, \ldots, \mathbf{v}_n\}$ be a linearly independent set of vectors in V. What condition should a linear transformation $T \colon V \to W$ satisfy so that $\{T\mathbf{v}_1, \ldots, T\mathbf{v}_n\}$ is a linearly independent subset of W?

(g) Let $\{\mathbf{u}_1, \ldots, \mathbf{u}_n\}$ be a basis for a subspace U of V. What conditions should a linear transformation $T \colon V \to W$ satisfy so that $\{T\mathbf{u}_1, \ldots, T\mathbf{u}_n\}$ is a basis for the subspace $T(U)$?

(h) Suppose the vectors $\mathbf{v}_1, \ldots, \mathbf{v}_n$ span the vector space V and $T \colon V \to W$ is a linear transformation. If $\{T\mathbf{v}_1, \ldots, T\mathbf{v}_n\}$ is a basis for W what can you conclude about the vectors $\mathbf{v}_1, \ldots, \mathbf{v}_n$? What can you conclude about the linear transformation T?

(i) Is it true that two finite dimensional vector spaces are isomorphic if and only if they have the same dimension?

(5) A sequence of vector spaces and linear maps

$$\cdots \longrightarrow V_{n-1} \xrightarrow{\ j_n\ } V_n \xrightarrow{\ j_{n+1}\ } V_{n+1} \longrightarrow \cdots$$

is said to be EXACT AT V_n if $\operatorname{ran} j_n = \ker j_{n+1}$. A sequence is EXACT if it is exact at each of its constituent vector spaces. A sequence of vector spaces and linear maps of the form

$$\mathbf{0} \longrightarrow U \xrightarrow{\ j\ } V \xrightarrow{\ k\ } W \longrightarrow \mathbf{0} \tag{1}$$

is a SHORT EXACT SEQUENCE. (Here $\mathbf{0}$ denotes the trivial 0-dimensional vector space, and the unlabeled arrows are the obvious linear maps.)

(a) The sequence (1) is exact at U if and only if j is injective.

(b) The sequence (1) is exact at W if and only if k is surjective.

(c) Let U and V be vector spaces. Then the following sequence is short exact:

$$0 \longrightarrow U \xrightarrow{\;\iota_1\;} U \oplus V \xrightarrow{\;\pi_2\;} V \longrightarrow 0.$$

The indicated linear maps are defined by

$$\iota_1 \colon U \to U \oplus V \colon a \mapsto (a, 0)$$

and

$$\pi_2 \colon U \oplus V \to V \colon (a, b) \mapsto b.$$

(d) Suppose $a < b$. Let $\mathcal{K}$ be the family of constant functions on the interval $[a, b]$, $\mathcal{C}^1$ be the family of all continuously differentiable functions on $[a, b]$, and $\mathcal{C}$ be the family of all continuous functions on $[a, b]$. Specify linear maps j and k so that the following sequence is short exact:

$$0 \longrightarrow \mathcal{K} \xrightarrow{\;j\;} \mathcal{C}^1 \xrightarrow{\;k\;} \mathcal{C} \longrightarrow 0.$$

(e) Let $\mathcal{C}$ be the family of all continuous functions on the interval $[0, 2]$. Let E_1 be the mapping from $\mathcal{C}$ into $\mathbb{R}$ defined by $E_1(f) = f(1)$. (The functional E_1 is called "evaluation at 1".)

Find a subspace $\mathcal{F}$ of $\mathcal{C}$ such that the following sequence is short exact.

$$0 \longrightarrow \mathcal{F} \xrightarrow{\;\iota\;} \mathcal{C} \xrightarrow{\;E_1\;} \mathbb{R} \longrightarrow 0.$$

(f) Suppose that the following sequence of finite dimensional vector spaces and linear maps is exact.

$$0 \longrightarrow V_n \xrightarrow{\;f_n\;} V_{n-1} \xrightarrow{\;f_{n-1}\;} \cdots \xrightarrow{\;f_2\;} V_1 \xrightarrow{\;f_1\;} V_0 \longrightarrow 0$$

Show that

$$\sum_{k=0}^{n} (-1)^k \dim(V_k) = 0.$$

Definition 8.3.1. It is frequently useful to think of functions as arrows in diagrams. For example, the situation $f\colon U \to X$, $g\colon X \to V$, $h\colon U \to W$, $j\colon W \to V$ may be represented by the following diagram.

$$
\begin{array}{ccc}
U & \xrightarrow{\ f\ } & X \\
\downarrow{\scriptstyle h} & & \downarrow{\scriptstyle g} \\
W & \xrightarrow{\ j\ } & V
\end{array}
$$

The diagram is said to COMMUTE (or to be a COMMUTATIVE DIAGRAM) if $j \circ h = g \circ f$.

(g) Suppose that in the following diagram of vector spaces and linear maps

$$
\begin{array}{ccccccccc}
0 & \longrightarrow & U & \xrightarrow{\ j\ } & V & \xrightarrow{\ k\ } & W & \longrightarrow & 0 \\
& & \downarrow{\scriptstyle f} & & \downarrow{\scriptstyle g} & & \vdownarrow{\scriptstyle h} & & \\
0 & \longrightarrow & U' & \xrightarrow[j']{} & V' & \xrightarrow[k']{} & W' & \longrightarrow & 0
\end{array}
$$

the rows are exact and the left square commutes. Then there exists a unique linear map $h\colon W \to W'$ that makes the right square commute.

In parts (h)–(k) consider the diagram

$$
\begin{array}{ccccccccc}
0 & \longrightarrow & U & \xrightarrow{\ j\ } & V & \xrightarrow{\ k\ } & W & \longrightarrow & 0 \\
& & \downarrow{\scriptstyle f} & & \downarrow{\scriptstyle g} & & \downarrow{\scriptstyle h} & & \\
0 & \longrightarrow & U' & \xrightarrow[j']{} & V' & \xrightarrow[k']{} & W' & \longrightarrow & 0
\end{array}
$$

where the rows are exact and the squares commute.

(h) If g is surjective, so is h.

(i) If f is surjective and g is injective, then h is injective.

(j) If f and h are surjective, so is g.

(k) If f and h are injective, so is g.

(6) Let V and W be vector spaces. Prove that (under the usual pointwise operations of addition and scalar multiplication) $\mathcal{L}(V, W)$ is a vector space.

(7) Let M be a subspace of a vector space V and V/M be the quotient space defined in Definition 5.1.7 in Chapter 5. Then the map

$$\pi \colon V \to V/M \colon x \mapsto [x]$$

is linear and is called the QUOTIENT MAP.

Remark 8.3.2. Imagine a vector space V of large dimension to which no coordinates have been assigned. Let M be an arbitrary subspace of V. Does M have a complementary subspace? Well, yes, in general, many of them. But how would we go about choosing one? Of course we could start by working with bases, and putting in coordinates, and writing equations, and making some arbitrary choices to specify some complementary space. But all that seems rather unnatural involving a lot of wasted energy. A better solution is to relax the demand that the "complement" be literally a subspace of V, and replace it with something very simple that for most purposes works just as well. That object is the quotient space V/M. No fuss, no muss, no bother. That is the point of the next problem.

(8) If M and N are complementary subspaces of a vector space V, then $N \cong V/M$. (So $V \cong M \oplus V/M$.) *Hint.* Consider the map $n \to [n]$ for every $n \in N$.

The following result is called the *fundamental quotient theorem*, or the *first isomorphism theorem*, for vector spaces.

(9) Prove the following theorem and its corollary.

Theorem 8.3.3. *Let V and W be vector spaces and $M \preceq V$. If $T \in \mathcal{L}(V, W)$ and $\ker T \supseteq M$, then there exists a unique $\widetilde{T} \in \mathcal{L}(V/M, W)$ that makes the following diagram commute.*

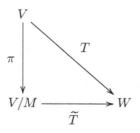

Furthermore, $\widetilde{T}$ is injective if and only if $\ker T = M$; and $\widetilde{T}$ is surjective if and only if T is.

Corollary 8.3.4. *If* $T\colon V \to W$ *is a linear map between vector spaces, then*

$$\operatorname{ran} T \cong V/\ker T.$$

8.4 Answers to Odd-Numbered Exercises

(1) (a) $-2, -1, 5, 5$
   (b) $6, 3, -2$

(3) line, $-\dfrac{x}{5} = \dfrac{y}{3} = z$, plane, $2x + y - z = 0$

(5) (a) $7, 8, 5$
   (b) $4, 21, 10$
   (c) $3, 14, 5$
   (d) I
   (e) 2

(7) Pa, DK, H, DK, Pa, H, DO, VA, DL, Pb, VA, H, VA, DK, H, VA, DO, H

(9) $\sin t$, $\cos t$, $y'' + y = 0$

(11) (a) line, $\dfrac{x}{4} = -\dfrac{3y}{2} = z$
   (b) plane, $2x - y - z = 0$

(13) $\mathbb{R}^4$ is the span of a, b, c, and d, all of which T takes to $\mathbf{0}$; so were T linear, its range would contain only $\mathbf{0}$.

(15) T does not map $\mathbf{0}$ to $\mathbf{0}$.

(17) If T were linear, then $T(-2, -2)$ would be $-2T(1, 1) = -2(3, -6) = (-6, 12)$.

Chapter 9

LINEAR MAPS BETWEEN EUCLIDEAN SPACES

9.1 Background

Topics: linear mappings between finite dimensional spaces, a matrix as a linear map, the representation of a linear map as a matrix.

Proposition 9.1.1. *Let $T \in \mathfrak{L}(V, W)$ where V is an n-dimensional vector space and W is an m-dimensional vector space and let $\{\mathbf{e}^1, \mathbf{e}^2, \ldots, \mathbf{e}^n\}$ be a basis for V. Define an $m \times n$-matrix $[T]$ whose k^{th} column $(1 \leq k \leq n)$ is the column vector $T\mathbf{e}^k$. Then for each $x \in V$ we have*

$$Tx = [T]x.$$

The displayed equation above requires a little interpretation. The left side is T evaluated at x; the right side is an $m \times n$ matrix multiplied by an $n \times 1$ matrix (that is, a column vector). Then the asserted equality can be thought of as identifying

(1) two vectors in $\mathbb{R}^m$,
(2) two m-tuples of real numbers, or
(3) two column vectors of length m (that is, two $m \times 1$ matrices).

If we wished to distinguish rigorously between column vectors and row vectors and also wished to identify m-tuples with row vectors, then the equation in the preceding proposition would have to read

$$Tx = ([T](x^t))^t.$$

To avoid the extra notation in these notes we will not make this distinction. In an equation interpret a vector as a row vector or as a column vector in any way that makes sense.

Definition 9.1.2. If V and W are finite dimensional vector spaces with bases and $T \in \mathcal{L}(V, W)$, then the matrix $[T]$ in the preceding proposition is the MATRIX REPRESENTATION of T. It is also called the STANDARD MATRIX for T. The RANK of an $m \times n$ matrix is the rank of the operator on $T \colon \mathbb{R}^n \to \mathbb{R}^m$ it represents (with respect to the standard bases on $\mathbb{R}^n$ and $\mathbb{R}^m$).

9.2　Exercises

(1) Let $T \colon \mathbb{R}^4 \to \mathbb{R}^3$ be defined by

$$T\mathbf{x} = (x_1 - 3x_3 + x_4, 2x_1 + x_2 + x_3 + x_4, 3x_2 - 4x_3 + 7x_4)$$

for every $\mathbf{x} = (x_1, x_2, x_3, x_4) \in \mathbb{R}^4$. (The map T is linear, but you need not prove this.)

(a) Find $[T]$. Answer: $\begin{bmatrix} & & \\ & & \\ & & \end{bmatrix}$.

(b) Find $T(1, -2, 1, 3)$. Answer: _____.

(c) Find $([T]((1, -2, 1, 3)^t))^t$. Answer: _____.

(d) Find $\ker T$. Answer: $\ker T = \text{span}\{$_____$\}$.

(e) Find $\operatorname{ran} T$. Answer: $\operatorname{ran} T = $ _____.

(2) Let $T \colon \mathbb{R}^3 \to \mathbb{R}^4$ be defined by

$$T\mathbf{x} = (x_1 - 3x_3, x_1 + x_2 - 6x_3, x_2 - 3x_3, x_1 - 3x_3)$$

for every $\mathbf{x} = (x_1, x_2, x_3) \in \mathbb{R}^3$. (The map T is linear, but you need not prove this.) Then

(a) $[T] =$ 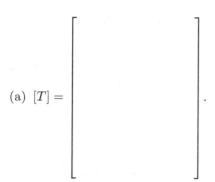 .

(b) $T(3, -2, 4) = $ _____.

(c) $\ker T = \text{span}\{$_____$\}$.

(d) $\text{ran}\, T = \text{span}\{$_____$\}$.

(3) Let $\mathcal{P}_n$ be the vector space of all polynomial functions on $\mathbb{R}$ with degree strictly less than n. The usual basis for $\mathcal{P}_n$ is the set of polynomials $1, t, t^2, t^3, \ldots, t^{n-1}$. Define $T \colon \mathcal{P}_3 \to \mathcal{P}_5$ by

$$Tf(x) = \int_0^x \int_0^u p(t)\, dt$$

for all $x,\, u \in \mathbb{R}$.

(a) Then the matrix representation of the linear map T with respect

to the usual bases for $\mathcal{P}_3$ and $\mathcal{P}_5$ is

$$\begin{bmatrix} & & \\ & & \\ & & \\ & & \\ & & \end{bmatrix}.$$

(b) The kernel of T is _____.

(c) The range of T is $\text{span}\{$_____$\}$.

(4) Let $\mathcal{P}_4$ be the vector space of polynomials of degree strictly less than 4. Consider the linear transformation $D^2 \colon \mathcal{P}_4 \to \mathcal{P}_4 \colon f \mapsto f''$.

(a) Then the matrix representation $[D^2]$ of D^2 (with respect to the usual

basis $\{1, t, t^2, t^3\}$ for $\mathcal{P}_4$) is given by $[D^2] = \begin{bmatrix} & & \\ & & \\ & & \\ & & \end{bmatrix}$.

(b) $\ker D^2 = \text{span}\{\underline{\hspace{2cm}}\}$.
(c) $\operatorname{ran} D^2 = \text{span}\{\underline{\hspace{2cm}}\}$.

(5) Let $\mathcal{P}_4$ be the vector space of polynomials of degree strictly less than 4 and $T\colon \mathcal{P}_4 \to \mathcal{P}_5$ be the linear transformation defined by $(Tp)(t) = (2 + 3t)p(t)$ for every $p \in \mathcal{P}_4$ and $t \in \mathbb{R}$. Then the matrix representation of T (with respect to the usual basis $\{1, t, t^2, t^3\}$ for $\mathcal{P}_4$) is given by

$$[T] = \begin{bmatrix} & & \\ & & \\ & & \\ & & \\ & & \end{bmatrix}.$$

(6) Let $T\colon \mathbb{R}^3 \to \mathbb{R}^3$ be the linear transformation whose standard matrix is $\begin{bmatrix} 1 & 1 & 0 \\ 0 & 1 & 1 \\ 1 & 0 & -1 \end{bmatrix}$. We know that T is not onto because the only vectors (u, v, w) that are in the range of T are those that satisfy the relation $u + av + bw = 0$ where $a = \underline{\hspace{1cm}}$ and $b = \underline{\hspace{1cm}}$.

(7) Let T be the linear map from $\mathbb{R}^3$ to $\mathbb{R}^3$ defined by

$$T(x, y, z) = (3x + 2y + z, x + 3z, -y + 4z).$$

(a) The matrix representation of T is given by

$$[T] = \begin{bmatrix} & & \\ & & \\ & & \end{bmatrix}.$$

(b) The range of T is (geometrically speaking) a $\underline{\hspace{2cm}}$ whose equation is $\underline{\hspace{4cm}}$.

9.3 Problems

(1) Define $T\colon \mathbb{R}^3 \to \mathbb{R}^2$ by

$$T\mathbf{x} = (x_1 + 2x_2 - x_3, x_2 + x_3)$$

for all $\mathbf{x} = (x_1, x_2, x_3)$ in $\mathbb{R}^3$.

(a) Explain how to find $[T]$, the matrix representation for T.
(b) Explain how to use $[T]$ to find $T(\mathbf{x})$ when $\mathbf{x} = (-1, 2, -1)$.
(c) Explain how to use $[T]$ to find a vector $\mathbf{x}$ in $\mathbb{R}^3$ such that $T\mathbf{x} = (0, 1)$.

Carry out the computations you describe.

(2) Let $\mathcal{P}$ be the vector space of all polynomial functions on $\mathbb{R}$ with real coefficients. Define linear transformations $T, D\colon \mathcal{P} \to \mathcal{P}$ by

$$(Dp)(x) = p'(x)$$

and

$$(Tp)(x) = x^2 p(x)$$

for all $x \in \mathbb{R}$.
Explain carefully how to find matrix representations for the linear transformations $D + T$, DT, and TD (with respect to the usual basis $\{1, t, t^2\}$ for the space of polynomials of degree two or less). Carry out the computations you describe. Use the resulting matrices to find $((D + T)(p))(x)$, $(DTp)(x)$, and $(TDp)(x)$ where $p(x) = 3x^2 + 4x - 3$ for all $x \in \mathbb{R}$.

(3) Define $T\colon \mathbb{R}^2 \to \mathbb{R}^3$ by

$$T\mathbf{x} = (x_1 - x_2, x_2 - x_1, x_1 + x_2)$$

for all $\mathbf{x} = (x_1, x_2)$ in $\mathbb{R}^2$.

(a) Explain carefully how to find $[T]$, the matrix representation for T.
(b) How do we use $[T]$ to find $T(1, -2)$?
(c) Are there any nonzero vectors $\mathbf{x}$ in $\mathbb{R}^2$ such that $T\mathbf{x} = \mathbf{0}$? Explain.
(d) Under what conditions is a vector (u, v, w) in the range of T? Explain.

(4) Let $\mathcal{C}^1([0, 1])$ be the vector space of all functions defined on the interval $[0, 1]$ that have continuous derivatives at each point and $\mathcal{C}([0, 1])$ be

the vector space of continuous functions on $[0, 1]$. Define a function $T: C^1([0, 1]) \to C([0, 1])$ by

$$(Tf)(x) = \int_0^x f(t)\, dt + f'(x)$$

for every $x \in [0, 1]$.

(a) Prove that the function T is linear.
(b) Let $f(x) = \sin x$ and $g(x) = \cos x$ for all $x \in [0, 1]$. Explain why one of these functions belongs to the kernel of T while the other does not.

(5) Let $\mathcal{P}_4$ be the vector space of polynomials of degree strictly less than 4. Consider the linear transformation $D^2: \mathcal{P}_4 \to \mathcal{P}_4: f \mapsto f''$. Explain carefully how to find $[D^2]$, the matrix representation of D^2 (with respect to the usual basis $\{1, t, t^2, t^3\}$ for $\mathcal{P}_4$). Then explain how to use $[D^2]$ to find $\ker D^2$ and $\operatorname{ran} D^2$.

9.4   Answers to Odd-Numbered Exercises

(1) (a) $\begin{bmatrix} 1 & 0 & -3 & 1 \\ 2 & 1 & 1 & 1 \\ 0 & 3 & -4 & 7 \end{bmatrix}$

(b) $(1, 4, 11)$

(c) $(1, 4, 11)$ (or $[1\ 4\ 11]$)

(d) $(1, -9, 2, 5)$

(e) $\mathbb{R}^3$

(3) (a) $\begin{bmatrix} 0 & 0 & 0 \\ 0 & 0 & 0 \\ \frac{1}{2} & 0 & 0 \\ 0 & \frac{1}{6} & 0 \\ 0 & 0 & \frac{1}{12} \end{bmatrix}$

(b) $\{0\}$

(c) x^2, x^3, x^4

(5) $\begin{bmatrix} 2 & 0 & 0 & 0 \\ 3 & 2 & 0 & 0 \\ 0 & 3 & 2 & 0 \\ 0 & 0 & 3 & 2 \\ 0 & 0 & 0 & 3 \end{bmatrix}$

(7) (a) $\begin{bmatrix} 3 & 2 & 1 \\ 1 & 0 & 3 \\ 0 & -1 & 4 \end{bmatrix}$

(b) plane, $u - 3v + 2w = 0$

Chapter 10

PROJECTION OPERATORS

10.1 Background

Topics: projections along one subspace onto another.

Definition 10.1.1. Let V be a vector space and suppose that $V = M \oplus N$. We know that for each $\mathbf{v} \in V$ there exist unique vectors $\mathbf{m} \in M$ and $\mathbf{n} \in N$ such that $\mathbf{v} = \mathbf{m} + \mathbf{n}$ (see Problem 5 in Chapter 5). Define a function $E_{MN} \colon V \to V$ by $E_{MN}v = n$. The function E_{MN} is the PROJECTION OF V ALONG M ONTO N. (Frequently we write E for E_{MN}. But keep in mind that E depends on both M and N.)

Theorem 10.1.2. *Let V be a vector space and suppose that $V = M \oplus N$. If E is the projection of V along M onto N, then*

(i) *E is linear;*
(ii) *$E^2 = E$ (that is, E is* IDEMPOTENT*);*
(iii) *$\operatorname{ran} E = N$; and*
(iv) *$\ker E = M$.*

Theorem 10.1.3. *Let V be a vector space and suppose that $E \colon V \to V$ is a function which satisfies*

(i) *E is linear, and*
(ii) *$E^2 = E$.*

Then

$$V = \ker E \oplus \operatorname{ran} E$$

and E is the projection of V along $\ker E$ onto $\operatorname{ran} E$.

Theorem 10.1.4. *Let V be a vector space and suppose that $V = M \oplus N$. If E is the projection of V along M onto N, then $I - E$ is the projection of V along N onto M.*

10.2   Exercises

(1) Let M be the line $y = 2x$ and N be the y-axis in $\mathbb{R}^2$. Then

$$[E_{MN}] = \begin{bmatrix} a & a \\ -b & c \end{bmatrix} \quad \text{and} \quad [E_{NM}] = \begin{bmatrix} c & a \\ b & a \end{bmatrix}$$

where $a =$ ___, $b =$ ___, and $c =$ ___.

(2) Let P be the plane in $\mathbb{R}^3$ whose equation is $x - z = 0$ and L be the line whose equations are $y = 0$ and $x = -z$. Let E be the projection of $\mathbb{R}^3$ along L onto P and F be the projection of $\mathbb{R}^3$ along P onto L. Then

$$[E] = \begin{bmatrix} a & b & a \\ b & c & b \\ a & b & a \end{bmatrix} \quad \text{and} \quad [F] = \begin{bmatrix} a & b & -a \\ b & b & b \\ -a & b & a \end{bmatrix}$$

where $a =$ ___, $b =$ ___, and $c =$ ___.

(3) Let P be the plane in $\mathbb{R}^3$ whose equation is $x + 2y - z = 0$ and L be the line whose equations are $\dfrac{x}{3} = y = \dfrac{z}{2}$. Let E be the projection of $\mathbb{R}^3$ along L onto P and F be the projection of $\mathbb{R}^3$ along P onto L. Then

$$[E] = \frac{1}{3} \begin{bmatrix} a & -b & c \\ -d & d & d \\ a-2d & -b+2d & c+2d \end{bmatrix} \quad \text{and} \quad [F] = \frac{1}{3} \begin{bmatrix} 3d & 3e & -3d \\ d & e & -d \\ 2d & 2e & -2d \end{bmatrix}$$

where $a =$ ___, $b =$ ___, $c =$ ___, $d =$ ___, and $e =$ ___.

(4) Let P be the plane in $\mathbb{R}^3$ whose equation is $x - y - 2z = 0$ and L be the line whose equations are $x = 0$ and $y = -z$. Let E be the projection of $\mathbb{R}^3$ along L onto P and F be the projection of $\mathbb{R}^3$ along P onto L. Then

$$[E] = \begin{bmatrix} a & b & b \\ -a & c & c \\ a & -a & -a \end{bmatrix} \quad \text{and} \quad [F] = \begin{bmatrix} b & b & b \\ a & -a & -c \\ -a & a & c \end{bmatrix}$$

where $a =$ ___, $b =$ ___, and $c =$ ___.

(5) Let E be the projection of $\mathbb{R}^3$ along the z-axis onto the plane $3x - y + 2z = 0$ and let F be the projection of $\mathbb{R}^3$ along the plane $3x - y + 2z = 0$ onto the z-axis.

(a) Then $[E] = \begin{bmatrix} & & \\ & & \\ & & \end{bmatrix}$.

(b) Where does F take the point $(4, 5, 1)$? Answer: (___, ___, ___).

(6) Let M be the y-axis and N be the plane $x + y - 2z = 0$ in $\mathbb{R}^3$.

(a) Then the projection E_{MN} of $\mathbb{R}^3$ along M onto N is
$\begin{bmatrix} & & \\ & & \\ & & \end{bmatrix}$.

(b) The projection E_{NM} takes the vector $(3, 2, 1)$ to (___, ___, ___).

10.3 Problems

(1) Let E be a projection on a vector space. Show that a vector $\mathbf{x}$ belongs to the range of E if and only if $E\mathbf{x} = \mathbf{x}$. *Hint.* Recall (from Theorems 10.1.2 and 10.1.3) that a projection is a linear map E such that $E^2 = E$.

(2) Prove Theorem 10.1.2.

(3) Prove Theorem 10.1.3.

(4) Prove Theorem 10.1.4. *Hint.* Use Theorem 10.1.3.

(5) Let P be the plane in $\mathbb{R}^3$ whose equation is $x - z = 0$ and L be the line whose equations are $y = 0$ and $x = \frac{1}{2}z$. Explain carefully how to find the matrix representation of the operator E_{LP}, that is, the projection of $\mathbb{R}^3$ along L onto P. Carry out the computation you describe.

(6) Let L be the line in $\mathbb{R}^3$ whose equations are $x = y$ and $z = 0$, and let P be the plane whose equation is $x - z = 0$. Explain carefully how to find the matrix representation of the operator E_{LP}, that is, the projection of $\mathbb{R}^3$ along L onto P. Carry out the computation you describe.

(7) Let P be the plane in $\mathbb{R}^3$ whose equation is $x - 3y + z = 0$ and L be the line whose equations are $x = -2y = -z$. Explain carefully how to

find the matrix representation of the operator E_{LP} of $\mathbb{R}^3$ along L onto P and the projection E_{PL} of $\mathbb{R}^3$ along P onto L.

(8) Prove that a linear transformation between vector spaces has a left inverse if and only if it is injective.

(9) Prove that a linear transformation between vector spaces has a right inverse if and only if it is surjective.

10.4   Answers to Odd-Numbered Exercises

(1) 0, 2, 1

(3) 0, 6, 3, 1, 2

(5) (a) $\begin{bmatrix} 1 & 0 & 0 \\ 0 & 1 & 0 \\ -\frac{3}{2} & \frac{1}{2} & 0 \end{bmatrix}$

(b) $0, 0, \frac{9}{2}$

Part 4

SPECTRAL THEORY OF VECTOR SPACES

Chapter 11

EIGENVALUES AND EIGENVECTORS

11.1 Background

Topics: characteristic polynomial, eigenvalues, eigenvectors, eigenspaces, algebraic multiplicity and geometric multiplicity of an eigenvalue.

Definition 11.1.1. A number λ is an EIGENVALUE of an operator T on a vector space V if $\ker(\lambda I_V - T)$ contains a nonzero vector. Any such vector is an EIGENVECTOR of T associated with λ and $\ker(\lambda I_V - T)$ is the EIGENSPACE of T associated with λ. The set of all eigenvalues of the operator T is its (POINT) SPECTRUM and is denoted by $\sigma(T)$.

If M is an $n \times n$ matrix, then $\det(\lambda I_n - M)$ (where I_n is the $n \times n$ identity matrix) is a polynomial in λ of degree n. This is the CHARACTERISTIC POLYNOMIAL of M. A standard way of computing the eigenvalues of an operator T on a finite dimensional vector space is to find the zeros of the characteristic polynomial of its matrix representation. It is an easy consequence of the multiplicative property of the determinant function (see Proposition 2.1.10) that the characteristic polynomial of an operator T on a vector space V is independent of the basis chosen for V and hence of the particular matrix representation of T that is used.

Theorem 11.1.2 (Spectral Mapping Theorem). *If T is an operator on a finite dimensional vector space and p is a polynomial, then*

$$\sigma(p(T)) = p(\sigma(T)).$$

That is, if $\sigma(T) = \{\lambda_1, \ldots, \lambda_k\}$, then $\sigma(p(T)) = \{p(\lambda_1), \ldots, p(\lambda_k)\}$.

11.2   Exercises

(1) If $A = \begin{bmatrix} 1 & -1 & 4 \\ 3 & 2 & -1 \\ 2 & 1 & -1 \end{bmatrix}$, then the eigenvalue _____ has corresponding eigenvector (_____, 1, 1), the eigenvalue _____ has corresponding eigenvector (_____, 4, 1), and the eigenvalue _____ has corresponding eigenvector (_____, 2, 1).

(2) Let $A = \begin{bmatrix} 0 & 0 & 2 \\ 0 & 2 & 0 \\ 2 & 0 & 0 \end{bmatrix}$.

   (a) The eigenvalues of A are _____, _____, and _____.
   (b) The matrix A has a one-dimensional eigenspace.
       It is the span of the vector $(1,$ _____, _____$)$.
   (c) The matrix A has a two-dimensional eigenspace.
       It is the span of the vectors $(1, 0,$ _____$)$ and $(0, 1,$ _____$)$.

(3) Choose a, b and c in the matrix $A = \begin{bmatrix} 0 & 1 & 0 \\ 0 & 0 & 1 \\ a & b & c \end{bmatrix}$ so that the characteristic polynomial of A is $-\lambda^3 + 4\lambda^2 + 5\lambda + 6$.

   Answer: $a =$ _____; $b =$ _____; and $c =$ _____.

(4) Suppose that it is known that the matrix $A = \begin{bmatrix} 1 & 0 & -1 \\ \sqrt{3} & a & 17 \\ 2 & 0 & b \end{bmatrix}$ has eigenvalues 2 and 3 and that the eigenvalue 2 has algebraic multiplicity 2. Then $a =$ _____ and $b =$ _____.

(5) The matrices $A = \begin{bmatrix} a & 1 \\ -2 & d \end{bmatrix}$ and $B = \frac{1}{25}\begin{bmatrix} 114 & 48 \\ 48 & 86 \end{bmatrix}$ have the same eigenvalues. Then $a =$ _____ and $d =$ _____.

(6) Let $A = \begin{bmatrix} 3 & 4 & 2 \\ 0 & 1 & 2 \\ 0 & 0 & 0 \end{bmatrix}$.

   (a) The eigenvalues of A are _____, _____, and _____.
   (b) The matrix A has three one-dimensional eigenspaces. They are spanned by the vectors (_____, _____, _____), (_____, _____, _____), and (_____, _____, _____), respectively.

(7) Let $A = \begin{bmatrix} 1 & 1 & 1 & 1 \\ 1 & 1 & 1 & 1 \\ 1 & 1 & 1 & 1 \\ 1 & 1 & 1 & 1 \end{bmatrix}$.

   (a) The eigenvalues of $A - I$ are ____ (which has algebraic multiplicity ____) and ____ (which has algebraic multiplicity ____).

   (b) The determinant of $A - I$ is ____.

(8) Let T be the operator on $\mathbb{R}^3$ whose matrix representation is $\begin{bmatrix} 1 & -1 & 0 \\ 0 & 0 & 0 \\ -2 & 2 & 2 \end{bmatrix}$. Then the eigenvalues of the operator $T^5 - 3T^4 + T^3 - T^2 + T - 3I$ are ____, ____, and ____.

(9) If A is a square matrix and $A^2 = I$ then its possible eigenvalues are ____, and ____.

11.3 Problems

(1) Suppose that A and B are $n \times n$ matrices. Prove that $\sigma(AB) = \sigma(BA)$. *Hint.* Show that if λ is an eigenvalue of AB, then it is also an eigenvalue of BA. Deal with the cases $\lambda = 0$ and $\lambda \neq 0$ separately.

(2) Let $c \in \mathbb{R}$. Suppose that A is an $n \times n$ matrix and that the sum of the entries in each column of A is c. Prove that c is an eigenvalue of A. *Hint.* Consider the sum of the row vectors of the matrix $A - cI$.

(3) Prove that every square matrix is the sum of two nonsingular matrices. *Hint.* For the matrix A use $A - cI$ and cI for some appropriate number c.

(4) This is a problem in cryptography. Read about Hill ciphers, then decode the following Hill 3-cipher given that the first two words of the plaintext are known to be "My candle". (See for example [1], Section 11.16.)

    OGWGCGWGKK.EWVD.XZJOHZWLNYH USTFAIOS.A.KBN
    JRCENYQZV,IE LTGCGWGKC YYBLSDWWODLBVFFOS.H

In many discussions of Hill ciphers letters of the alphabet are assigned numbers from 0 to 25 and arithmetic is done modulo 26. The encoding here is done slightly differently. Here each letter is assigned its numerical position in the alphabet (including Z that is assigned 26). Furthermore, a space between words is assigned 27, a comma is assigned 28, and a period is assigned zero. Thus, for this code, all arithmetic should be done modulo 29. (One reason for this is that some computer algebra systems have problems calculating inverses mod 26.) Note: the ciphertext contains exactly three spaces.

11.4   Answers to Odd-Numbered Exercises

(1) $-2, -1, 1, -1, 3, 1$

(3) $6, 5, 4$

(5) $2, 6$ (or $6, 2$)

(7) (a) $-1, 3, 3, 1$
    (b) -3

Chapter 12

DIAGONALIZATION
OF MATRICES

12.1 Background

Topics: similarity of matrices, triangular and diagonal matrices, diagonalization, annihilating and minimal polynomials, algebraic and geometric multiplicity of an eigenvalue, the *Cayley-Hamilton theorem*.

Definition 12.1.1. Two operators R and T on a vector space V are SIMILAR if there exists an invertible operator S on V such that $R = S^{-1}TS$.

Proposition 12.1.2. *If V is a vector space, then similarity is an equivalence relation on $\mathfrak{L}(V)$.*

Definition 12.1.3. Let V be a finite dimensional vector space and $B = \{e^1, \ldots, e^n\}$ be a basis for V. An operator T on V is DIAGONAL if there exist scalars $\alpha_1, \ldots, \alpha_n$ such that $Te^k = \alpha_k e^k$ for each $k \in \mathbb{N}_n$. Equivalently, T is diagonal if its matrix representation $[T] = [t_{ij}]$ has the property that $t_{ij} = 0$ whenever $i \neq j$.

Asking whether a particular operator on some finite dimensional vector space is diagonal is, strictly speaking, nonsense. As defined the operator property of being diagonal is definitely *not* a vector space concept. It makes sense only for a vector space *for which a basis has been specified*. This important, if obvious, fact seems to go unnoticed in many beginning linear algebra texts, due, I suppose, to a rather obsessive fixation on $\mathbb{R}^n$ in such courses. Here is the relevant *vector space* property.

Definition 12.1.4. An operator T on a finite dimensional vector space V is DIAGONALIZABLE if there exists a basis for V with respect to which T is diagonal. Equivalently, an operator on a finite dimensional vector space *with basis* is diagonalizable if it is similar to a diagonal operator. If a matrix D is diagonalizable and $\Lambda = S^{-1}DS$ is diagonal, we say that the matrix S DIAGONALIZES D.

Theorem 12.1.5. *Let A be an $n \times n$ matrix with n linear independent eigenvectors. If S is a matrix with these eigenvectors as columns, then S diagonalizes A. The entries along the diagonal of the resulting diagonal matrix are all eigenvalues of A.*

Definition 12.1.6. A polynomial is MONIC if its leading coefficient is 1. Thus a polynomial of degree n is monic if it takes the form $x^n + a_{n-1}x^{n-1} + \cdots + a_1 x + a_0$.

Definition 12.1.7. Let p be a polynomial of degree at least one and T be an operator on some vector space. We say that p is an ANNIHILATING POLYNOMIAL for T (or that p ANNIHILATES T) if $p(T) = 0$. For example, if $T^3 - 4T^2 + T - 7I = 0$, then the polynomial p defined by $p(x) = x^3 - 4x^2 + x - 7$ is an annihilating polynomial for T.

Definition 12.1.8. Let T be an operator on a finite dimensional vector space. The MINIMAL POLYNOMIAL of T is the unique monic polynomial of smallest degree which annihilates T. (It is left as a problem to verify the existence and the uniqueness of such a polynomial: see problem 8.)

Theorem 12.1.9 (Cayley-Hamilton Theorem). *On a finite dimensional vector space the characteristic polynomial of an operator T annihilates T.*

Paraphrase: Every matrix satisfies its characteristic equation.

Definition 12.1.10. A square matrix $A = [a_{ij}]$ is UPPER TRIANGULAR if $a_{ij} = 0$ whenever $i > j$. A matrix is TRIANGULABLE (or TRIANGULARIZABLE) if it is similar to an upper triangular matrix.

Theorem 12.1.11. *Let T be an operator on a finite dimensional vector space and let $\{\lambda_1, \ldots, \lambda_k\}$ be its distinct eigenvalues. Then:*

(1) *T is triangulable if and only if its minimal polynomial can be factored into a product of linear factors. That is, if and only if there are positive*

integers $r_1, \ldots, r_k$ *such that*

$$m_T(x) = (x - \lambda_1)^{r_1} \cdots (x - \lambda_k)^{r_k}.$$

(2) T *is diagonalizable if and only if its minimal polynomial has the form*

$$m_T(x) = (x - \lambda_1) \cdots (x - \lambda_k).$$

Corollary 12.1.12. *Every operator on a complex finite dimensional vector space is triangulable.*

Definition 12.1.13. An operator is NILPOTENT if some power of the operator is **0**.

12.2 Exercises

(1) Let $A = \begin{bmatrix} 1 & 1 & 1 \\ 1 & 1 & 1 \\ 1 & 1 & 1 \end{bmatrix}$. The characteristic polynomial of A is $\lambda^p (\lambda - 3)^q$
where $p =$ _____ and $q =$ _____.
The minimal polynomial of A is $\lambda^r (\lambda - 3)^s$ where $r =$ _____ and $s =$ _____.

(2) Let T be the operator on $\mathbb{R}^4$ whose matrix representation is
$$\begin{bmatrix} 0 & 1 & 0 & -1 \\ -2 & 3 & 0 & -1 \\ -2 & 1 & 2 & -1 \\ 2 & -1 & 0 & 3 \end{bmatrix}.$$
The characteristic polynomial of T is $(\lambda - 2)^p$ where $p =$ _____.
The minimal polynomial of T is $(\lambda - 2)^r$ where $r =$ _____.

(3) Let T be the operator on $\mathbb{R}^3$ whose matrix representation is
$$\begin{bmatrix} 3 & 1 & -1 \\ 2 & 2 & -1 \\ 2 & 2 & 0 \end{bmatrix}.$$

(a) Find the characteristic polynomial of T.
Answer: $c_T(\lambda) = (\lambda - 1)^p (\lambda - 2)^q$ where $p =$ _____ and $q =$ _____.
(b) Find the minimal polynomial of T.
Answer: $m_T(\lambda) = (\lambda - 1)^r (\lambda - 2)^s$ where $r =$ _____ and $s =$ _____.

(c) Find the eigenspaces M_1 and M_2 of T.

Answer: $M1 = \text{span}\{$_____$\}$ and $M_2 = \text{span}\{$_____$\}$.

(4) Let T be the operator on $\mathbb{R}^5$ whose matrix representation is
$$\begin{bmatrix} 1 & 0 & 0 & 1 & -1 \\ 0 & 1 & -2 & 3 & -3 \\ 0 & 0 & -1 & 2 & -2 \\ 1 & -1 & 1 & 0 & 1 \\ 1 & -1 & 1 & -1 & 2 \end{bmatrix}.$$

(a) Find the characteristic polynomial of T.

Answer: $c_T(\lambda) = (\lambda+1)^p(\lambda-1)^q$ where $p =$ _____ and $q =$ _____.

(b) Find the minimal polynomial of T.

Answer: $m_T(\lambda) = (\lambda+1)^r(\lambda-1)^s$ where $r =$ _____ and $s =$ _____.

(c) Find the eigenspaces V_1 and V_2 of T.

Answer: $V_1 = \text{span}\{$_____$\}$ and

$V_2 = \text{span}\{$_____$\}$.

(5) Let T be an operator whose matrix representation is
$$\begin{bmatrix} 0 & 0 & 0 & 0 & 0 \\ 0 & 0 & 0 & 0 & 0 \\ 3 & 1 & 0 & 0 & 0 \\ 0 & 0 & 0 & 1 & 2 \\ 0 & 0 & 0 & -1 & -1 \end{bmatrix}.$$

(a) Regarding T as an operator on $\mathbb{R}^5$ find its characteristic polynomial and minimal polynomial.

Answer: $c_T(\lambda) = \lambda^p(\lambda^2+1)^q$ where $p =$ _____ and $q =$ _____.

and $m_T(\lambda) = \lambda^r(\lambda^2+1)^s$ where $r =$ _____ and $s =$ _____.

(b) Regarded as an operator on $\mathbb{R}^5$ is T diagonalizable? _____. Is it triangulable? _____.

(c) Regarded as an operator on $\mathbb{C}^5$ is T diagonalizable? _____. Is it triangulable? _____.

(6) Let T be the operator on $\mathbb{R}^3$ whose matrix representation is
$$\begin{bmatrix} 2 & 0 & 0 \\ -1 & 3 & 2 \\ 1 & -1 & 0 \end{bmatrix}.$$

(a) Find the characteristic and minimal polynomials of T.

Answer: $c_T(\lambda) = (\lambda - 1)^p(\lambda - 2)^q$ where $p =$ _____ and $q =$ _____.

and $m_T(\lambda) = (\lambda - 1)^r(\lambda - 2)^s$ where $r =$ _____ and $s =$ _____.

(b) What can be concluded from the form of the minimal polynomial?

Answer: _____.

(c) Find a matrix S (if one exists) that diagonalizes $[T]$. What is the diagonal form Λ of $[T]$ produced by this matrix? Answer:

$$S = \begin{bmatrix} a & b & a \\ b & b & -c \\ -b & a & b \end{bmatrix} \text{ where } a = \underline{\quad}, \ b = \underline{\quad}, \text{ and } c = \underline{\quad}; \text{ and}$$

$$\Lambda = \begin{bmatrix} \lambda & 0 & 0 \\ 0 & \mu & 0 \\ 0 & 0 & \mu \end{bmatrix} \text{ where } \lambda = \underline{\quad} \text{ and } \mu = \underline{\quad}.$$

(7) Let T be the operator on $\mathbb{R}^3$ whose matrix representation is
$$\begin{bmatrix} 8 & -6 & 12 \\ -18 & 11 & 18 \\ -6 & -3 & 26 \end{bmatrix}.$$

(a) Find the characteristic and minimal polynomials of T.

Answer: $c_T(\lambda) = (\lambda - 5)^p(\lambda - 20)^q$ where $p =$ _____ and $q =$ _____.

and $m_T(\lambda) = (\lambda - 5)^r(\lambda - 20)^s$ where $r =$ _____ and $s =$ _____.

(b) What can be concluded from the form of the minimal polynomial?

Answer: _____.

(c) Find a matrix S (if one exists) that diagonalizes $[T]$. What is the diagonal form Λ of $[T]$ produced by this matrix? Answer: $S =$

$$\begin{bmatrix} a & b & c \\ d & -a & a \\ b & c & b \end{bmatrix} \text{ where } a = \underline{\quad}, \ b = \underline{\quad}, \ c = \underline{\quad}, \text{ and } d = \underline{\quad}; \text{ and}$$

$$\Lambda = \begin{bmatrix} \lambda & 0 & 0 \\ 0 & \mu & 0 \\ 0 & 0 & \mu \end{bmatrix} \text{ where } \lambda = \underline{\quad} \text{ and } \mu = \underline{\quad}.$$

(8) Let $\mathcal{P}_n$ be the space of polynomials of degree strictly less than n and D be the differentiation operator on $\mathcal{P}_n$. Then

(a) the only eigenvalue of D is $\lambda =$ _____;

(b) the corresponding eigenspace is the span of _____;

(c) the algebraic multiplicity of λ is _____; and

(d) the geometric multiplicity of λ is _____.

(9) Suppose that A is a 2×2 matrix of real numbers, that $A^2 = I$, and that A is neither the identity matrix I nor $-I$. Then $\operatorname{tr} A =$ _____ and $\det A =$ _____.

(10) Suppose that A is as in the preceding exercise and that the entries in its first row are 3 and -1 (in that order). Then the entries in the second row are _____ and _____.

12.3 Problems

(1) Prove that the trace function is a similarity invariant on the family of $n \times n$ matrices; that is, prove that if A and B are similar $n \times n$ matrices, then $\operatorname{tr} A = \operatorname{tr} B$. *Hint.* Prove first that if M and N are $n \times n$ matrices, then MN and NM have the same trace.

(2) Prove that the determinant function is a similarity invariant on the family of $n \times n$ matrices; that is, prove that if A and B are similar $n \times n$ matrices, then $\det A = \det B$.

(3) Prove that if two matrices are diagonalized by the same matrix, then they commute.

(4) Prove that if a matrix A is diagonalizable, then so is every matrix similar to A.

(5) Show that if A is a diagonalizable $n \times n$ matrix of real (or complex) numbers, then $\operatorname{tr} A$ is the sum of the eigenvalues of A and $\det A$ is their product.

(6) Suppose that T is an operator on a finite dimensional complex vector space and that $\sigma(T) = \{0\}$. Show that T is nilpotent.

(7) Let T be an operator on a finite dimensional vector space.

   (a) Show that if p is an annihilating polynomial for T, then the minimal polynomial m_T divides p. *Hint.* Suppose that p annihilates T (so that the degree of p is at least as large as the degree of m_T). Divide p by m_T. Then there exist polynomials q (the quotient) and r (the remainder) such that

$$p = m_T q + r \quad \text{and} \quad \text{degree of } r < \text{degree of } m_T.$$

Conclude that $r = 0$.

(b) Show that the minimal polynomial m_T and the characteristic polynomial c_T have exactly the same roots (although not necessarily the same multiplicities). *Hint.* To show that every root of m_T is also a root of c_T, it is enough to know that m_T divides c_T. Why is that true?

To obtain the converse, suppose that λ is a root of c_T: that is, suppose that λ is an eigenvalue of T. Use the *spectral mapping theorem* 11.1.2 to show that $m_T(\lambda) = 0$.

(8) Let T be an operator on a finite dimensional vector space V. Show that there is a unique monic polynomial of smallest degree which annihilates T. *Hint.* This asks for a proof of the existence and the uniqueness of the *minimal polynomial* for the operator T. The existence part is easy: If there are any polynomials at all which annihilate T, surely there is one of smallest degree. (How do we know that there is at least one polynomial that annihilates T?) We want the annihilating polynomial of smallest degree to be monic — is this a problem?

To verify the uniqueness of the minimal polynomial, consider the case of degree one separately. That is, suppose that p and q are monic annihilating polynomials of degree one. How do we know that $p = q$? Then consider polynomials of higher degree. Suppose that p and q are monic annihilating polynomials of smallest degree k where $k > 1$. What can you say about $p - q$?

12.4   Answers to Odd-Numbered Exercises

(1) 2, 1, 1, 1

(3) (a) 1, 2
    (b) 1, 2
    (c) $(1, 0, 2)$, $(1, 1, 2)$

(5) (a) 3, 1, 2, 1
    (b) no, no
    (c) no, yes

(7) (a) 1, 2, 1, 1
    (b) T is diagonalizable
    (c) 2, 1, 0, 3, 5, 20

(9) 0, -1

Chapter 13

SPECTRAL THEOREM
FOR VECTOR SPACES

13.1 Background

Topics: the *spectral theorem for finite dimensional vector spaces* (writing a diagonalizable operator as a linear combination of projections).

The central fact asserted by the finite dimensional vector space version of the *spectral theorem* is that every diagonalizable operator on such a space can be written as a linear combination of projection operators where the coefficients of the linear combination are the eigenvalues of the operator and the ranges of the projections are the corresponding eigenspaces. Here is a formal statement of the theorem.

Theorem 13.1.1 (Spectral theorem: vector space version). *Let T be a diagonalizable operator on a finite dimensional vector space V, and $\lambda_1, \ldots, \lambda_k$ be the (distinct) eigenvalues of T. For each j let M_j be the eigenspace associated with λ_j and E_j be the projection of V onto M_j along $M_1 + \cdots + M_{j-1} + M_{j+1} + \cdots + M_k$. Then*

(i) $T = \lambda_1 E_1 + \cdots + \lambda_k E_k$,

(ii) $I = E_1 + \cdots + E_k$, *and*

(iii) $E_i E_j = 0$ *when* $i \neq j$.

For a proof of this result see, for example, [6], page 215, Theorem 11.

13.2   Exercises

(1) Let T be the operator on $\mathbb{R}^2$ whose matrix representation is $\begin{bmatrix} -7 & 8 \\ -16 & 17 \end{bmatrix}$.

(a) Find the characteristic polynomial and minimal polynomial for T.
Answer: $c_T(\lambda) = $ _____ and $m_T(\lambda) = $
_____.

(b) The eigenspace M_1 associated with the smaller eigenvalue λ_1 is the span of $(1, \underline{\quad})$.

(c) The eigenspace M_2 associated with the larger eigenvalue λ_2 is the span of $(1, \underline{\quad})$.

(d) We wish to write T as a linear combination of projection operators. Find the (matrix representations of the) appropriate projections E_1 and E_2 onto the eigenspaces M_1 and M_2, respectively.

Answer: $E_1 = \begin{bmatrix} a & b \\ a & b \end{bmatrix}$, where $a = \underline{\quad}$ and $b = \underline{\quad}$, and $E_2 = \begin{bmatrix} -c & c \\ -d & d \end{bmatrix}$, where $c = \underline{\quad}$ and $d = \underline{\quad}$.

(e) Compute the sum and product of E_1 and E_2.

Answer: $E_1 + E_2 = \begin{bmatrix} & \\ & \end{bmatrix}$; and $E_1 E_2 = \begin{bmatrix} & \\ & \end{bmatrix}$.

(f) Write T as a linear combination of the projections found in (d).
Answer: $[T] = \underline{\quad} E_1 + \underline{\quad} E_2$.

(g) Find a matrix S that diagonalizes $[T]$. What is the associated diagonal form Λ of $[T]$?

Answer: $S = \begin{bmatrix} 1 & 1 \\ a & b \end{bmatrix}$, where $a = \underline{\quad}$ and $b = \underline{\quad}$, and $\Lambda = \begin{bmatrix} & \\ & \end{bmatrix}$.

(2) Let T be the operator on $\mathbb{R}^3$ whose matrix representation is

$$\begin{bmatrix} -2 & -1 & -6 \\ -6 & -1 & -12 \\ 2 & 1 & 6 \end{bmatrix}.$$

(a) Find the characteristic polynomial and minimal polynomial for T.

Answer: $c_T(\lambda) = $ _____ and $m_T(\lambda) = $ _____.

(b) The eigenspace M_1 associated with the smallest eigenvalue λ_1 is the span of $(3, \text{\_\_\_}, \text{\_\_\_})$.

(c) The eigenspace M_2 associated with the middle eigenvalue λ_2 is the span of $(1, \text{\_\_\_}, \text{\_\_\_})$.

(d) The eigenspace M_3 associated with the largest eigenvalue λ_3 is the span of $(1, \text{\_\_\_}, \text{\_\_\_})$.

(e) We wish to write T as a linear combination of projection operators. Find the (matrix representations of the) appropriate projections E_1, E_2, and E_3 onto the eigenspaces M_1, M_2, and M_3, respectively.

Answer: $E_1 = \begin{bmatrix} a & c & a \\ 2a & c & 2a \\ -b & c & -b \end{bmatrix}$; $E_2 = \begin{bmatrix} -b & d & c \\ -2a & a & c \\ b & -d & c \end{bmatrix}$; and $E_3 = \begin{bmatrix} c & -d & -a \\ c & -b & -2a \\ c & d & a \end{bmatrix}$, where $a = $ ___, $b = $ ___, $c = $ ___, and $d = $ ___.

(f) Write T as a linear combination of the projections found in (e).

Answer: $[T] = $ _____ $E_1 + $ _____ $E_2 + $ _____ E_3.

(g) Find a matrix S that diagonalizes $[T]$. What is the associated diagonal form Λ of $[T]$?

Answer: $S = \begin{bmatrix} a & b & b \\ 2a & a & c \\ -c & -b & -b \end{bmatrix}$, where $a = $ ___, $b = $ ___, and $c = $ ___, and $\Lambda = \begin{bmatrix} & & \\ & & \\ & & \end{bmatrix}$.

(3) Find a matrix A whose eigenvalues are 1 and 4, and whose eigenvectors are $(3, 1)$ and $(2, 1)$, respectively.

Answer: $A = \begin{bmatrix} & \\ & \end{bmatrix}$.

(4) Let T be the operator on $\mathbb{R}^3$ whose matrix representation is
$$\begin{bmatrix} 2 & -2 & 1 \\ -1 & 1 & 1 \\ -1 & 2 & 0 \end{bmatrix}.$$

(a) Write T as a linear combination of projections.

Answer: $T = c_1 E_1 + c_2 E_2 + c_3 E_3$ where $c_1 = $ _____, $c_2 = $ _____, $c_3 = $ _____,

$$E_1 = \begin{bmatrix} a & b & -b \\ a & b & -b \\ a & -b & b \end{bmatrix}, E_2 = \begin{bmatrix} b & a & b \\ b & a & b \\ b & a & b \end{bmatrix}, \text{ and } E_3 = \begin{bmatrix} b & -b & a \\ -b & b & a \\ -b & b & a \end{bmatrix}$$

where $a = $ ____ and $b = $ ____.

(b) Calculate the following:

$$E_1 E_2 = \begin{bmatrix} & & \\ & & \\ & & \end{bmatrix} ; E_1 E_3 = \begin{bmatrix} & & \\ & & \\ & & \end{bmatrix} ; E_2 E_3 = $$

$$\begin{bmatrix} & & \\ & & \\ & & \end{bmatrix}.$$

(c) $E_1 + E_2 + E_3 = \begin{bmatrix} & & \\ & & \\ & & \end{bmatrix}.$

(d) Write T^3 as a linear combination of E_1, E_2, and E_3.

Answer: $T^3 = $ _____ $E_1 + $ _____ $E_2 + $ _____ E_3.

(5) Let $A = \begin{bmatrix} 1 & 0 & 0 \\ 3 & 4 & -3 \\ 0 & 0 & 1 \end{bmatrix}.$

(a) Find the eigenvalues of A. Answer: The smaller eigenvalue is $\lambda_1 = $ _____; and the larger eigenvalue is $\lambda_2 = $ _____.

(b) The eigenspace associated with λ_1 is span$\{(1, 0, \_\_\_), (1, \_\_\_, 0)\}$; and the eigenspace associated with λ_2 is span$\{(\_\_\_, 1, 0)\}$.

(c) Find a factorization of A in the form $S\Lambda S^{-1}$ where Λ is a diagonal matrix.

Answer: $S = \begin{bmatrix} 1 & a & b \\ 0 & -a & a \\ a & b & b \end{bmatrix}$; $\Lambda = \begin{bmatrix} c & 0 & 0 \\ 0 & c & 0 \\ 0 & 0 & d \end{bmatrix}$; $S^{-1} = \begin{bmatrix} j & j & k \\ k & j & -k \\ k & k & -k \end{bmatrix}$; where $a = \underline{\quad}$, $b = \underline{\quad}$, $c = \underline{\quad}$, $d = \underline{\quad}$, $j = \underline{\quad}$, and $k = \underline{\quad}$.

(d) Find a square root of A.

Answer: $\sqrt{A} = \begin{bmatrix} a & b & b \\ a & c & -a \\ b & b & a \end{bmatrix}$; where $a = \underline{\quad}$, $b = \underline{\quad}$, and $c = \underline{\quad}$.

13.3 Problem

(1) Explain carefully how to use spectral theory (the theory of eigenvalues and eigenvectors) to find a square root of the matrix $A = \begin{bmatrix} 10 & -6 \\ -6 & 10 \end{bmatrix}$. Illustrate your discussion by carrying out the computation.

13.4 Answers to Odd-Numbered Exercises

(1) (a) $\lambda^2 - 10\lambda + 9$, $\lambda^2 - 10\lambda + 9$

(b) 1

(c) 2

(d) $2, -1, 1, 2$

(e) $\begin{bmatrix} 1 & 0 \\ 0 & 1 \end{bmatrix}$, $\begin{bmatrix} 0 & 0 \\ 0 & 0 \end{bmatrix}$

(f) 1, 9

(g) $1, 2, \begin{bmatrix} 1 & 0 \\ 0 & 9 \end{bmatrix}$

(3) $\begin{bmatrix} -5 & 18 \\ -3 & 10 \end{bmatrix}$

(5) (a) 1, 4

    (b) 1, -1, 0

    (c) 1, 0, 1, 4, 0, 1

    (d) 1, 0, 2

Chapter 14

SOME APPLICATIONS OF THE SPECTRAL THEOREM

14.1 Background

Topics: systems of linear differential equations, initial conditions, steady state solutions, the functional calculus for operators on finite dimensional vector spaces, Markov processes.

Theorem 14.1.1. *Let*

$$\frac{d\mathbf{u}}{dt} = A\mathbf{u} \tag{14.1.1}$$

be a vector differential equation (that is, a system of ordinary linear differential equations) where A is an $n \times n$ matrix and suppose that $\mathbf{u}_0 = \mathbf{u}(0)$ is an initial value of the system. If A is a diagonalizable matrix (so that $A = S\Lambda S^{-1}$ for some diagonal matrix Λ and some invertible matrix S), then the equation (14.1.1) has the solution

$$\mathbf{u}(t) = e^{At}\mathbf{u}_0 = Se^{\Lambda t}S^{-1}\mathbf{u}_0.$$

Definition 14.1.2. A MARKOV MATRIX is a square matrix with nonnegative entries and with each column adding to 1.

Proposition 14.1.3 (Facts about Markov matrices.). *Let A be a Markov matrix. Then*

(i) *$\lambda_1 = 1$ is an eigenvalue.*
(ii) *The eigenvector $\mathbf{e}_1$ corresponding to λ_1 is nonnegative and it is a steady state.*

(iii) *The other eigenvalues satisfy* $|\lambda_k| \leq 1$.
(iv) *If any power of A has all entries strictly positive, then* $|\lambda_k| < 1$ *for all*
$k \neq 1$ *and* $A^k \mathbf{u}_0 \to \mathbf{u}_\infty$ *where the steady state* $\mathbf{u}_\infty$ *is a multiple of* $\mathbf{e}_1$.

14.2   Exercises

(1) Let $A = \begin{bmatrix} -1 & 1 \\ 1 & -1 \end{bmatrix}$.

(a) The eigenvalues of A are $\lambda_1 = $ _____ and $\lambda_2 = $ _____.

(b) The corresponding eigenvectors are $\mathbf{e}_1 = (1, a)$ and $\mathbf{e}_2 = (a, -a)$
where $a = $ ____.

(c) Then $e^{At} = a \begin{bmatrix} 1 + e^{-bt} & 1 - e^{-bt} \\ 1 - e^{-bt} & 1 + e^{-bt} \end{bmatrix}$ where $a = $ ____ and $b = $ ____.

(d) Let $\mathbf{u}(t) = (x(t), y(t))$. The general solution to the system of equations $\dfrac{d\mathbf{u}}{dt} = A\mathbf{u}$ with initial conditions $x_0 = 3$ and $y_0 = 1$ is
$x(t) = a + be^{-ct}$ and $y(t) = a - be^{-ct}$ where $a = $ ____, $b = $ ____, and
$c = $ ____.

(e) Find the steady state solution to the system $\dfrac{d\mathbf{u}}{dt} = A\mathbf{u}$ under the
initial conditions given in (d). That is, find $\mathbf{u}_\infty = \begin{bmatrix} x_\infty \\ y_\infty \end{bmatrix}$ where
$x_\infty = \lim_{t\to\infty} x(t)$ and $y_\infty = \lim_{t\to\infty} y(t)$.

Answer: $\mathbf{u}_\infty = \begin{bmatrix} \quad \\ \quad \end{bmatrix}$.

(2) Suppose that at time t the population $y(t)$ of a predator and the
population $x(t)$ of its prey are governed by the equations

$$\frac{dx}{dt} = 4x - 2y$$

$$\frac{dy}{dt} = x + y.$$

If at time $t = 0$ the populations are $x = 300$ and $y = 200$, then
the populations at all future times t are $x(t) = ae^{bt} + 200e^{ct}$ and
$y(t) = de^{bt} + ae^{ct}$ where $a = $ _____, $b = $ ____, $c = $ ____, and $d = $ _____.
The long run ratio of populations of prey to predator approaches _____
to _____.

(3) Use matrix methods to solve the initial value problem consisting of the system of differential equations

$$\frac{du}{dt} = 4u - v - w$$

$$\frac{dv}{dt} = u + 2v - w$$

$$\frac{dw}{dt} = u - v + 2w$$

and the initial conditions

$$u(0) = 2 \qquad v(0) = -2 \qquad w(0) = 7.$$

Answer: $u(t) = ae^{bt} - e^{at}$; $v(t) = ae^{bt} - ce^{at}$; and $w(t) = ae^{bt} + de^{at}$ where $a = $ ____, $b = $ ____, $c = $ ____, and $d = $ ____.

(4) Consider the initial value problem: $y'' - y' - 2y = 0$ with the initial conditions $y_0 = 3$ and $y_0' = 3$.

(a) Express the differential equation in the form $\dfrac{d\mathbf{u}}{dt} = A\mathbf{u}$ where $\mathbf{u} = (y, z)$ and $z = y'$. Then A is the matrix $\begin{bmatrix} a & b \\ c & b \end{bmatrix}$ where $a = $ ____, $b = $ ____, and $c = $ ____.

(b) The smaller eigenvalue of A is ____ and the larger is ____. The corresponding eigenvectors are $(1, a)$ and $(1, b)$ where $a = $ ____ and $b = $ ____.

(c) The diagonal form of A is $\Lambda = \begin{bmatrix} a & 0 \\ 0 & b \end{bmatrix}$ where $a = $ ____ and $b = $ ____.

(d) Find the diagonalizing matrix S for A. That is, find S so that $\Lambda = S^{-1}AS$. Answer: $S = \begin{bmatrix} 1 & 1 \\ a & b \end{bmatrix}$ where $a = $ ____ and $b = $ ____.

(e) The matrix e^{At} is $\dfrac{1}{a}\begin{bmatrix} be^{ct} + e^{dt} & -e^{ct} + e^{dt} \\ -be^{ct} + be^{dt} & e^{ct} + be^{dt} \end{bmatrix}$ where $a = $ ____, $b = $ ____, $c = $ ____, and $d = $ ____.

(f) The solution to the initial value problem is $y(t) = $ _____.

(5) Use *the spectral theorem* to solve the initial value problem

$$y''' - 3y'' + 2y' = 0$$

where $y(0) = 2$, $y'(0) = 0$, and $y''(0) = 3$.

Answer: $y(t) = a + be^t + ce^{dt}$ where $a = $ _____, $b = $ _____, $c = $ _____, and $d = $ _____.

(6) Let $G_0 = 0$ and $G_1 = \dfrac{1}{2}$. For each $k \geq 0$ let G_{k+2} be the average of G_k and G_{k+1}.

    (a) Find the transition matrix A that takes the vector (G_{k+1}, G_k) to the vector (G_{k+2}, G_{k+1}).

        Answer: $A = \begin{bmatrix} & \\ & \end{bmatrix}$.

    (b) Find a diagonal matrix Λ that is similar to A.

        Answer: $\Lambda = \begin{bmatrix} & \\ & \end{bmatrix}$.

    (c) Find a matrix S such that $A = S\Lambda S^{-1}$.

        Answer: $S = \begin{bmatrix} & \\ & \end{bmatrix}$.

    (d) Determine the long run behavior of the numbers G_k.
        Answer: $G_\infty := \lim_{k \to \infty} G_k = $ _____.

(7) Let T be the operator on $\mathbb{R}^2$ whose matrix representation is $\begin{bmatrix} -7 & 8 \\ -16 & 17 \end{bmatrix}$. Use the *spectral theorem* to find $\sqrt{T}$. (A *square root* of T is an operator whose square is T.)
Answer: $\sqrt{T} = \begin{bmatrix} -1 & a \\ b & c \end{bmatrix}$ where $a = $ ____, $b = $ ____, and $c = $ ____.

(8) Let $A = \begin{bmatrix} 4 & 3 \\ 1 & 2 \end{bmatrix}$. Find A^{100}. (Write an *exact* answer — not a decimal approximation.)

    Answer: $A^{100} = \dfrac{1}{4} \begin{bmatrix} & \\ & \end{bmatrix}$.

(9) Let T be the operator on $\mathbb{R}^3$ whose matrix representation is
$$\begin{bmatrix} 2 & -2 & 1 \\ -1 & 1 & 1 \\ -1 & 2 & 0 \end{bmatrix}.$$

(a) Write T as a linear combination of projections.

Answer: $T = c_1 E_1 + c_2 E_2 + c_3 E_3$ where $c_1 =$ ____, $c_2 =$ ____,

$c_3 =$ ____,

$$E_1 = \begin{bmatrix} a & b & -b \\ a & b & -b \\ a & -b & b \end{bmatrix}, \quad E_2 = \begin{bmatrix} b & a & b \\ b & a & b \\ b & a & b \end{bmatrix}, \quad \text{and } E_3 = \begin{bmatrix} b & -b & a \\ -b & b & a \\ -b & b & a \end{bmatrix}$$

where $a =$ ____ and $b =$ ____.

(b) Calculate the following: $E_1 E_2 = \begin{bmatrix} & & \\ & & \\ & & \end{bmatrix}$; $E_1 E_3 =$

$$\begin{bmatrix} & & \\ & & \\ & & \end{bmatrix}; \quad E_2 E_3 = \begin{bmatrix} & & \\ & & \\ & & \end{bmatrix}.$$

(c) $E_1 + E_2 + E_3 = \begin{bmatrix} & & \\ & & \\ & & \end{bmatrix}.$

(d) Write T^3 as a linear combination of E_1, E_2, and E_3.

Answer: $T^3 =$ ____ $E_1 +$ ____ $E_2 +$ ____ E_3.

(10) Let A be the matrix whose eigenvalues are $\lambda_1 = -1$, $\lambda_2 = 1/2$,

and $\lambda_3 = 1/3$, and whose corresponding eigenvectors are $\mathbf{e}_1 = \begin{bmatrix} 1 \\ 0 \\ 1 \end{bmatrix}$,

$\mathbf{e}_2 = \begin{bmatrix} 1 \\ -1 \\ 0 \end{bmatrix}$, and $\mathbf{e}_3 = \begin{bmatrix} 0 \\ -1 \\ 0 \end{bmatrix}$.

(a) Solve the difference equation $\mathbf{x}_{k+1} = A\mathbf{x}_k$ (where $\mathbf{x}_k = \begin{bmatrix} u_k \\ v_k \\ w_k \end{bmatrix}$)
subject to the initial condition $\mathbf{x}_0 = \begin{bmatrix} 10 \\ 20 \\ 30 \end{bmatrix}$.

Answer: $u_k = a(-b)^k - cd^k$, $v_k = cd^k$, and $w_k = a(-b)^k$ where
$a = $ ___, $b = $ ___, $c = $ ___, and $d = $ ___.

(b) Each $\mathbf{x}_k$ can be written as a linear combination of the vectors
(___, ___, ___) and (___, ___, ___).

(c) The value of $\mathbf{x}_{1000}$ is approximately (___, ___, ___).

(11) Let A be as in the preceding exercise. Solve the differential equa-
tion $\dfrac{d\mathbf{x}}{dt} = A\mathbf{x}$ subject to the initial conditions $\mathbf{x}_0 = \begin{bmatrix} 10 \\ 20 \\ 30 \end{bmatrix}$. Answer:
$\mathbf{x}(t) = \left(ae^{-t} - be^{ct}, be^{ct}, ae^{-t} \right)$ where $a = $ ___, $b = $ ___, and $c = $ ___.

(12) Suppose three cities A, B, and C are centers for trucks. Every month
half of the trucks in A and half of those in B go to C. The other half
of the trucks in A and the other half of the trucks in B stay where
they are. Every month half of the trucks in C go to A and the other
half go to B.

(a) What is the (Markov) transition matrix that acts on the vector
$\begin{bmatrix} a_0 \\ b_0 \\ c_0 \end{bmatrix}$ (where a_0 is the number of trucks initially in A, etc.)?

Answer: $\begin{bmatrix} & & \\ & & \\ & & \end{bmatrix}$.

(b) If there are always 450 trucks, what is the long run distribution
of trucks? Answer: $a_\infty = $ _____, $b_\infty = $ _____, $c_\infty = $ _____.

14.3   Problems

(1) Initially a 2100 gallon tank M is full of water and an 1800 gallon tank
N is full of water in which 100 pounds of salt has been dissolved. Fresh
water is pumped into tank M at a rate of 420 gallons per minute and

salt water is released from tank N at the same rate. Additionally, the contents of M are pumped into N at a rate of 490 gallons per minute and the contents of N are pumped into M at a rate sufficient to keep both tanks full.

How long does it take (to the nearest second) for the concentration of salt in tank M to reach a maximum? And how much salt is there (to three significant figures) in M at that time?

(2) Show that if A is a diagonalizable $n \times n$ matrix, then

$$\det(\exp A) = \exp(\operatorname{tr} A).$$

Hint. What would be a reasonable definition of $\exp A$ if A were a diagonal matrix?

(3) Explain carefully how to use matrix methods to solve the initial value problem

$$y'' - y' - 6y = 0$$

under the initial conditions $y_0 = -2$ and $y_0' = 14$. Carry out the computations you describe.

(4) Explain carefully how to use matrix methods to solve the initial value problem consisting of the system of differential equations

$$\begin{cases} \dfrac{dv}{dt} = -v + w \\[2mm] \dfrac{dw}{dt} = v - w \end{cases} \qquad \text{and the initial conditions} \qquad \begin{cases} v(0) = 5 \\ w(0) = -3. \end{cases}$$

Carry out the computation you describe.

(5) Show how to use the *spectral theorem* to solve the initial value problem consisting of the system of differential equations

$$\frac{du}{dt} = -7u - 5v + 5w$$

$$\frac{dv}{dt} = 2u + 3v - 2w$$

$$\frac{dw}{dt} = -8u - 5v + 6w$$

and the initial conditions

$$u(0) = 2 \qquad v(0) = 1 \qquad w(0) = -1.$$

(6) Explain carefully how to use the *spectral theorem* to find a square root of the matrix $A = \begin{bmatrix} 2 & 1 \\ 1 & 2 \end{bmatrix}$. Illustrate your discussion by carrying out the computation.

(7) Let $A = \begin{bmatrix} 0 & 1 & 0 \\ 0 & 0 & 0 \\ 0 & 0 & 0 \end{bmatrix}$.

    (a) Does A have a cube root? Explain.

    (b) Does A have a square root? Explain.

(8) Let A be a symmetric 2×2 matrix whose trace is 20 and whose determinant is 64. Suppose that the eigenspace associated with the smaller eigenvalue of A is the line $x - y = 0$. Find a matrix B such that $B^2 = A$.

14.4   Answers to Odd-Numbered Exercises

(1) (a) 0, −2
    (b) 1
    (c) $\dfrac{1}{2}$, 2
    (d) 2, 1, 2
    (e) $\begin{bmatrix} 2 \\ 2 \end{bmatrix}$

(3) 3, 2, 5, 4

(5) $\dfrac{7}{2}$, −3, $\dfrac{3}{2}$, 2

(7) 2, −4, 5

(9) (a) −1, 1, 3, 0, $\dfrac{1}{2}$

    (b) $\begin{bmatrix} 0 & 0 & 0 \\ 0 & 0 & 0 \\ 0 & 0 & 0 \end{bmatrix}$, $\begin{bmatrix} 0 & 0 & 0 \\ 0 & 0 & 0 \\ 0 & 0 & 0 \end{bmatrix}$, $\begin{bmatrix} 0 & 0 & 0 \\ 0 & 0 & 0 \\ 0 & 0 & 0 \end{bmatrix}$

    (c) $\begin{bmatrix} 1 & 0 & 0 \\ 0 & 1 & 0 \\ 0 & 0 & 1 \end{bmatrix}$

    (d) −1, 1, 27

(11) 30, 20, $\dfrac{1}{2}$

Chapter 15

EVERY OPERATOR IS DIAGONALIZABLE PLUS NILPOTENT

15.1 Background

Topics: generalized eigenspaces, nilpotent operators

Definition 15.1.1. An operator T on a vector space is NILPOTENT if $T^n = 0$ for some $n \in \mathbb{N}$. Similarly, a square matrix is NILPOTENT if some power of it is the zero matrix.

Theorem 15.1.2. *Let T be an operator on a finite dimensional vector space V. Suppose that the minimal polynomial for T factors completely into linear factors*

$$m_T(x) = (x - \lambda_1)^{r_1} \cdots (x - \lambda_k)^{r_k}$$

where $\lambda_1, \ldots \lambda_k$ are the (distinct) eigenvalues of T. For each j let $W_j = \ker(T - \lambda_j I)^{r_j}$ and E_j be the projection of V onto W_j along $W_1 + \cdots + W_{j-1} + W_{j+1} + \cdots + W_k$. Then

$$V = W_1 \oplus W_2 \oplus \cdots \oplus W_k,$$

each W_j is invariant under T, and $I = E_1 + \cdots + E_k$. Furthermore, the operator

$$D = \lambda_1 E_1 + \cdots + \lambda_k E_k$$

is diagonalizable, the operator

$$N = T - D$$

is nilpotent, and N commutes with D.

A proof of this theorem can be found in [6], pages 220–223.

Corollary 15.1.3. *Every operator on a finite dimensional complex vector space is the sum of a diagonal operator and a nilpotent one.*

Definition 15.1.4. Since, in the preceding theorem, $T = D + N$ where D is diagonalizable and N is nilpotent, we say that D is the DIAGONALIZABLE PART of T and N is the NILPOTENT PART of T. The subspace $W_j = \ker(T - \lambda_j I)^{r_j}$ is GENERALIZED EIGENSPACE associated with the eigenvalue λ_j.

15.2   Exercises

(1) Let T be the operator on $\mathbb{R}^2$ whose matrix representation is $\begin{bmatrix} 2 & 1 \\ -1 & 4 \end{bmatrix}$.

   (a) Explain briefly why T is not diagonalizable.
      Answer: _____.

   (b) Find the diagonalizable and nilpotent parts of T. Answer: $D = \begin{bmatrix} a & b \\ b & a \end{bmatrix}$ and $N = \begin{bmatrix} -c & c \\ -c & c \end{bmatrix}$ where $a =$ ___, $b =$ ___, and $c =$ ___.

(2) Let T be the operator on $\mathbb{R}^2$ whose matrix representation is $\begin{bmatrix} 4 & -2 \\ 2 & 0 \end{bmatrix}$.

   (a) Explain briefly why T is not diagonalizable.
      Answer: _____.

   (b) Find the diagonalizable and nilpotent parts of T. Answer: $D = \begin{bmatrix} & \\ & \end{bmatrix}$ and $N = \begin{bmatrix} & \\ & \end{bmatrix}$.

(3) Let T be the operator on $\mathbb{R}^3$ whose matrix representation is $\begin{bmatrix} 1 & 1 & 0 \\ 0 & 1 & 0 \\ 0 & 0 & 0 \end{bmatrix}$.

   (a) The characteristic polynomial of T is $(\lambda)^p (\lambda - 1)^q$ where $p =$ ___ and $q =$ ___.

(b) The minimal polynomial of T is $(\lambda)^r(\lambda - 1)^s$ where $r =$ ___ and $s =$ ___.

(c) Explain briefly how we know from (b) that T is not diagonalizable.

Answer: _____.

(d) The eigenspace M_1 (corresponding to the smaller eigenvalue of T) is span$\{($___, ___, $1)\}$ and the eigenspace M_2 (corresponding to the larger eigenvalue) is span$\{(1,$ ___, ___$)\}$.

(e) Explain briefly how we know from (d) that T is not diagonalizable.

Answer: _____.

(f) The generalized eigenspace W_1 (corresponding to the smaller eigenvalue) is $W_1 =$ span$\{($___, ___, $1)\}$ and the generalized eigenspace W_2 (corresponding to the larger eigenvalue) is span$\{(1, a, a), (a, 1, a)\}$, where $a =$ ___.

(g) The (matrix representing the) projection E_1 of $\mathbb{R}^3$ along W_2 onto W_1 is $\begin{bmatrix} a & a & a \\ a & a & a \\ a & a & b \end{bmatrix}$ where $a =$ ___ and $b =$ ___.

(h) The (matrix representing the) projection E_2 of $\mathbb{R}^3$ along W_1 onto W_2 is $\begin{bmatrix} a & b & b \\ b & a & b \\ b & b & b \end{bmatrix}$ where $a =$ ___ and $b =$ ___.

(i) The diagonalizable part of T is $D = \begin{bmatrix} a & b & b \\ b & a & b \\ b & b & b \end{bmatrix}$ and the nilpotent part of T is $N = \begin{bmatrix} b & a & b \\ b & b & b \\ b & b & b \end{bmatrix}$ where $a =$ ___ and $b =$ ___.

(j) A matrix S that diagonalizes D is $\begin{bmatrix} a & b & a \\ a & a & b \\ b & a & a \end{bmatrix}$ where $a =$ ___ and $b =$ ___.

(k) The diagonal form Λ of the diagonalizable part D of T is $\begin{bmatrix} a & a & a \\ a & b & a \\ a & a & b \end{bmatrix}$ where $a =$ ___ and $b =$ ___.

(l) Show that D commutes with N by computing DN and ND.

Answer: $DN = ND = \begin{bmatrix} a & b & a \\ a & a & a \\ a & a & a \end{bmatrix}$ where $a =$ ___ and $b =$ ___ .

(4) Let T be the operator on $\mathbb{R}^3$ whose matrix representation is $\begin{bmatrix} 1 & 1 & 0 \\ 0 & 1 & 1 \\ 0 & 0 & 0 \end{bmatrix}$.

(a) The characteristic polynomial of T is $(\lambda)^p(\lambda - 1)^q$ where $p =$ ___ and $q =$ ___ .

(b) The minimal polynomial of T is $(\lambda)^r(\lambda - 1)^s$ where $r =$ ___ and $s =$ ___ .

(c) Explain briefly how we know from (b) that T is not diagonalizable.
Answer: _____ .

(d) The eigenspace M_1 (corresponding to the smaller eigenvalue of T) is span$\{(1,$ ___ , ___ $)\}$ and the eigenspace M_2 (corresponding to the larger eigenvalue) is span$\{(1,$ ___ , ___ $)\}$

(e) Explain briefly how we know from (d) that T is not diagonalizable.
Answer: _____ .

(f) The generalized eigenspace W_1 (corresponding to the smaller eigenvalue) is $W_1 =$ span$\{(1,$ ___ , ___ $)\}$ and the generalized eigenspace W_2 (corresponding to the larger eigenvalue) is span$\{(1, a, a), (a, 1, a)\}$, where $a =$ ___ .

(g) The (matrix representing the) projection E_1 of $\mathbb{R}^3$ along W_2 onto W_1 is $\begin{bmatrix} a & a & b \\ a & a & -b \\ a & a & b \end{bmatrix}$ where $a =$ ___ and $b =$ ___ .

(h) The (matrix representing the) projection E_2 of $\mathbb{R}^3$ along W_1 onto W_2 is $\begin{bmatrix} a & b & -a \\ b & a & a \\ b & b & b \end{bmatrix}$ where $a =$ ___ and $b =$ ___ .

(i) The diagonalizable part of T is $D = \begin{bmatrix} a & b & -a \\ b & a & a \\ b & b & b \end{bmatrix}$ and the nilpotent part of T is $N = \begin{bmatrix} b & a & a \\ b & b & b \\ b & b & b \end{bmatrix}$ where $a =$ ___ and $b =$ ___ .

(j) A matrix S that diagonalizes D is $\begin{bmatrix} a & a & b \\ -a & b & a \\ a & b & b \end{bmatrix}$ where $a =$ ___ and $b =$ ___.

(k) The diagonal form Λ of the diagonalizable part D of T is $\begin{bmatrix} a & a & a \\ a & b & a \\ a & a & b \end{bmatrix}$
where $a =$ ___ and $b =$ ___.

(l) When comparing this exercise with the preceding one it may seem that the correct answer to part (i) should be that the diagonalizable part of T is $D = \begin{bmatrix} 1 & 0 & 0 \\ 0 & 1 & 0 \\ 0 & 0 & 0 \end{bmatrix}$ and the nilpotent part of $[T]$ is $N = \begin{bmatrix} 0 & 1 & 0 \\ 0 & 0 & 1 \\ 0 & 0 & 0 \end{bmatrix}$, because D is diagonal, N is nilpotent, and $[T] = D + N$. Explain briefly why this is not correct.

Answer: _____.

(5) Let T be the operator on $\mathbb{R}^3$ whose matrix representation is $\begin{bmatrix} 3 & 1 & -1 \\ 2 & 2 & -1 \\ 2 & 2 & 0 \end{bmatrix}$.

(a) The characteristic polynomial of T is $(\lambda - 1)^p (\lambda - 2)^q$ where $p =$ ___ and $q =$ ___.

(b) The minimal polynomial of T is $(\lambda - 1)^r (\lambda - 2)^s$ where $r =$ ___ and $s =$ ___.

(c) The eigenspace M_1 (corresponding to the smaller eigenvalue of T) is span$\{(1, $ ___ , ___ $)\}$ and the eigenspace M_2 (corresponding to the larger eigenvalue) is span$\{(1, $ ___ , ___ $)\}$.

(d) The generalized eigenspace W_1 (corresponding to the smaller eigenvalue) is span$\{(1, $ ___ , ___ $)\}$ and the generalized eigenspace W_2 (corresponding to the larger eigenvalue) is span$\{(1, a, b), (0, b, a)\}$, where $a =$ ___ and $b =$ ___.

(e) The diagonalizable part of T is $D = \begin{bmatrix} a & a & b \\ b & c & b \\ -c & c & c \end{bmatrix}$ and the nilpotent

part of T is

$N = \begin{bmatrix} c & b & -a \\ c & b & -a \\ 2c & b & -c \end{bmatrix}$ where $a =$ ___, $b =$ ___, and $c =$ ___.

(f) A matrix S that diagonalizes D is $\begin{bmatrix} a & a & b \\ b & a & b \\ c & b & a \end{bmatrix}$ where $a =$ ___,

$b =$ ___, and $c =$ ___.

(g) The diagonal form Λ of the diagonalizable part D of T is

$\begin{bmatrix} \lambda & 0 & 0 \\ 0 & \mu & 0 \\ 0 & 0 & \mu \end{bmatrix}$ where $\lambda =$ ___ and $\mu =$ ___.

(h) Show that D commutes with N by computing DN and ND.

Answer: $DN = ND = \begin{bmatrix} a & b & -c \\ a & b & -c \\ 2a & b & -a \end{bmatrix}$ where $a =$ ___, $b =$ ___, and

$c =$ ___.

(6) Let T be the operator on $\mathbb{R}^4$ whose matrix representation is

$\begin{bmatrix} 0 & 1 & 0 & -1 \\ -2 & 3 & 0 & -1 \\ -2 & 1 & 2 & -1 \\ 2 & -1 & 0 & 3 \end{bmatrix}$.

(a) The characteristic polynomial of T is $(\lambda - 2)^p$ where $p =$ ___.

(b) The minimal polynomial of T is $(\lambda - 2)^r$ where $r =$ ___.

(c) The diagonalizable part of T is $D = \begin{bmatrix} a & b & b & b \\ b & a & b & b \\ b & b & a & b \\ b & b & b & a \end{bmatrix}$ where $a =$ ___

and $b =$ ___.

(d) The nilpotent part of T is $N = \begin{bmatrix} -a & b & c & -b \\ -a & b & c & -b \\ -a & b & c & -b \\ a & -b & c & b \end{bmatrix}$ where $a =$ ___,

$b =$ ___, and $c =$ ___.

(7) Let T be the operator on $\mathbb{R}^5$ whose matrix representation is

$$\begin{bmatrix} 1 & 0 & 0 & 1 & -1 \\ 0 & 1 & -2 & 3 & -3 \\ 0 & 0 & -1 & 2 & -2 \\ 1 & -1 & 1 & 0 & 1 \\ 1 & -1 & 1 & -1 & 2 \end{bmatrix}.$$

(a) Find the characteristic polynomial of T.

Answer: $c_T(\lambda) = (\lambda + 1)^p (\lambda - 1)^q$ where $p =$ ___ and $q =$ ___.

(b) Find the minimal polynomial of T.

Answer: $m_T(\lambda) = (\lambda + 1)^r (\lambda - 1)^s$ where $r =$ ___ and $s =$ ___.

(c) Find the eigenspaces M_1 and M_2 of T.

Answer: $M_1 = \text{span}\{(a, 1, b, a, a)\}$ where $a =$ ___ and $b =$ ___; and

$M_2 = \text{span}\{(1, a, b, b, b), (b, b, b, 1, a)\}$ where $a =$ ___ and $b =$ ___.

(d) Find the diagonalizable part of T.

Answer: $D = \begin{bmatrix} a & b & b & b & b \\ b & a & -c & c & -c \\ b & b & -a & c & -c \\ b & b & b & a & b \\ b & b & b & b & a \end{bmatrix}$ where $a =$ ___, $b =$ ___, and

$c =$ ___.

(e) Find the nilpotent part of T.

Answer: $N = \begin{bmatrix} a & a & a & b & -b \\ a & a & a & b & -b \\ a & a & a & a & a \\ b & -b & b & -b & b \\ b & -b & b & -b & b \end{bmatrix}$ where $a =$ ___ and $b =$ ___.

(f) Find a matrix S that diagonalizes the diagonalizable part D of T. What is the diagonal form Λ of D associated with this matrix?

Answer: $S = \begin{bmatrix} a & b & a & a & a \\ b & a & b & a & a \\ b & a & a & b & a \\ a & a & a & b & b \\ a & a & a & a & b \end{bmatrix}$ where $a =$ ___ and $b =$ ___.

$$\text{and } \Lambda = \begin{bmatrix} -a & 0 & 0 & 0 & 0 \\ 0 & a & 0 & 0 & 0 \\ 0 & 0 & a & 0 & 0 \\ 0 & 0 & 0 & a & 0 \\ 0 & 0 & 0 & 0 & a \end{bmatrix} \text{ where } a = \underline{\quad}.$$

15.3 Problems

(1) Explain in detail how to find the diagonalizable and nilpotent parts
of the matrix $A = \begin{bmatrix} -3 & -4 & 5 \\ 6 & 8 & -6 \\ -2 & -1 & 4 \end{bmatrix}$. Carry out the computations you
describe.

(2) Consider the matrix $A = \begin{bmatrix} 2 & 0 & -2 \\ 0 & 0 & 2 \\ 0 & 2 & 0 \end{bmatrix}$. In each part below explain
carefully what you are doing.

   (a) Find the characteristic polynomial for A.
   (b) Find the minimal polynomial for A. What can you conclude from
   the form of the minimal polynomial?
   (c) Find the eigenspace associated with each eigenvalue of A. Do the
   eigenvectors of A span $\mathbb{R}^3$? What can you conclude from this?
   (d) Find the generalized eigenspaces W_1 and W_2 associated with the
   eigenvalues of A.
   (e) Find the projection operators E_1 of $\mathbb{R}^3$ onto W_1 along W_2 and E_2
   of $\mathbb{R}^3$ onto W_2 along W_1.
   (f) Find the diagonalizable part D of A. Express D both as a single
   matrix and as a linear combination of projections.
   (g) Find a matrix S that diagonalizes D. What is the resulting diagonal
   form Λ of D?
   (h) Find the nilpotent part N of A. What is the smallest power of N
   that vanishes?

(3) Let T be the operator on $\mathbb{R}^3$ whose matrix representation is
$\begin{bmatrix} 1 & -1 & 0 \\ 1 & 3 & -1 \\ 0 & 0 & 1 \end{bmatrix}$.

   (a) Explain how to find the characteristic polynomial for T. Then carry
   out the computation.

(b) What is the *minimal polynomial* for a matrix? Find the minimal polynomial for T and explain how you know your answer is correct. What can you conclude from the form of this polynomial?

(c) Find the eigenspaces associated with each eigenvalue of T. Do the eigenvectors of T span $\mathbb{R}^3$? What can you conclude from this?

(d) Find the generalized eigenspaces W_1 and W_2 associated with the eigenvalues of T.

(e) Find the projection operators E_1 of $\mathbb{R}^3$ onto W_1 along W_2 and E_2 of $\mathbb{R}^3$ onto W_2 along W_1.

(f) Find the diagonalizable part D of T. Express D both as a single matrix and as a linear combination of projections.

(g) Find a matrix S that diagonalizes D. What is the resulting diagonal form Λ of D?

(h) Find the nilpotent part N of T. What is the smallest power of N that vanishes?

15.4   Answers to Odd-Numbered Exercises

(1) (a) The single one-dimensional eigenspace does not span $\mathbb{R}^2$. (OR: the minimal polynomial $(\lambda - 3)^2$ has a second degree factor — see theorem 12.1.11.)

(b) 3, 0, 1

(3) (a) 1, 2

(b) 1, 2

(c) The minimal polynomial has a second degree factor (see Theorem 12.1.11).

(d) 0, 0, 0, 0

(e) The eigenspaces do not span $\mathbb{R}^3$.

(f) 0, 0, 0

(g) 0, 1

(h) 1, 0

(i) 1, 0

(j) 0, 1

(k) 0, 1

(l) 0, 1

(5) (a) 1, 2

(b) 1, 2

(c) 0, 2, 1, 2

   (d) 0, 2, 1, 0
   (e) 1, 0, 2
   (f) 1, 0, 2
   (g) 1, 2
   (h) 4, 0, 2

(7) (a) 1, 4
   (b) 1, 2
   (c) 0, 1, 1, 0
   (d) 1, 0, 2
   (e) 0, 1
   (f) 0, 1, 1

Part 5

THE GEOMETRY OF
INNER PRODUCT SPACES

Chapter 16

COMPLEX ARITHMETIC

16.1 Background

Topics: complex arithmetic, absolute value and argument of a complex number, *De Moivre's theorem.*

Notation 16.1.1. Let a and b be real numbers and $z = a + bi$. Then $\operatorname{Re} z := a$ (the REAL PART of z) and $\operatorname{Im} z := b$ (the IMAGINARY PART of z).

16.2 Exercises

(1) $\operatorname{Re}\left(\dfrac{2+3i}{3-4i}\right) = $ _____.

(2) $\operatorname{Im}\left(\dfrac{2-3i}{2+3i}\right) = $ _____.

(3) $\left|\dfrac{21+7i}{1-2i}\right| = a\sqrt{b}$ where $a = $ ___ and $b = $ ___.

(4) $\operatorname{Arg}(-2\sqrt{3}+2i) = $ _____.

(5) Let $z = \dfrac{2 - \sqrt{3} - (1 + 2\sqrt{3})i}{1 + 2i}$. Then $z^{10} = a(1 - bi)$ where $a = $ _____ and $b = $ _____, $|z^{10}| = $ _____, and $\operatorname{Arg} z^{10} = $ _____.

(6) If $z = \dfrac{1}{\sqrt{2}}(1-i)$, then $z^{365} = a+bi$ where $a = $ _____ and $b = $ _____.

(7) The cube roots of -27 are $a + bi$, $a - bi$, and $c + di$ where $a = $ ____, $b = $ _____, $c = $ ____, and $d = $ ___.

(8) The complex number $w = 1 + 2i$ has 13 thirteenth roots. The sum of these roots is $a + bi$ where $a =$ _____ and $b =$ _____.

(9) Let $z_1 = (1, i, 1 + i)$, $z_2 = (i, 0, 1 - i)$, and $z_3 = (-1, 1 + 2i, 7 + 3i)$. Show that the set $\{z_1, z_2, z_3\}$ is linearly dependent in $\mathbb{C}^3$ by finding scalars α and β such that $\alpha z_1 + \beta z_2 - z_3 = 0$.

Answer: $\alpha =$ _____ and $\beta =$ _____.

(10) Let $A = \begin{bmatrix} 1 & 1 & 1 & i \\ 1 & i & 1 & i \\ 0 & 1+i & 0 & 0 \\ 2 & 0 & 2 & 2i \end{bmatrix}$. Then the rank of A is _____ and the nullity of A is _____.

(11) Let $A = \begin{bmatrix} 0 & 1 & 1 \\ 0 & i & 1 \\ i & i & i \end{bmatrix}$. Then $A^{-1} = \dfrac{1}{2} \begin{bmatrix} -2a & c & -2b \\ a+b & -a-b & c \\ a-b & a+b & c \end{bmatrix}$ where $a =$ _____, $b =$ _____, and $c =$ _____.

(12) If $A = \begin{bmatrix} i & 0 & -i \\ -2i & 1 & -1-4i \\ 2-i & i & 3 \end{bmatrix}$, then

$A^{-1} = \begin{bmatrix} -a - ai & b + ai & -a + bi \\ a - di & -c + ai & -a - di \\ -a + bi & b + ai & -a + bi \end{bmatrix}$ where $a =$ _____, $b =$ _____, $c =$ _____, and $d =$ _____.

(13) Consider the initial value problem: $y'' + 4y = 0$ with the initial conditions $y_0 = 3$ and $y'_0 = 2$.

    (a) Express the differential equation in the form $\dfrac{d\mathbf{u}}{dt} = A\mathbf{u}$ where $\mathbf{u} = (y, z)$ and $z = y'$. Then A is the matrix $\begin{bmatrix} a & b \\ -c & a \end{bmatrix}$ where $a =$ ___, $b =$ ___, and $c =$ ___.

    (b) The eigenvalues of A are _____ and _____; and the corresponding eigenvectors are $(1, a)$ and $(1, -a)$ where $a =$ _____.

    (c) The diagonal form of A is $\Lambda = \begin{bmatrix} a & 0 \\ 0 & -a \end{bmatrix}$ where $a =$ _____.

    (d) Find the diagonalizing matrix S for A. That is, find S so that $\Lambda = S^{-1}AS$. Answer: $S = \begin{bmatrix} 1 & 1 \\ a & -a \end{bmatrix}$ where $a =$ _____.

(e) Find the matrix e^{At}. Express your answer in terms of real trigonometric functions — *not* complex exponentials. Answer: $e^{At} =$
$$\begin{bmatrix} f(t) & a\,g(t) \\ -b\,g(t) & f(t) \end{bmatrix} \text{ where } a = \underline{\hspace{1cm}}, \ b = \underline{\hspace{1cm}}, \ f(t) = \underline{\hspace{2cm}},$$
and $g(t) = \underline{\hspace{2cm}}$.

(f) The solution to the initial value problem is $y(t) = \underline{\hspace{3cm}}$.

16.3 Problems

(1) Explain in detail how to find all the cube roots of i.

(2) Show that three points z_1, z_2, and z_3 in the complex plane form an
equilateral triangle if and only if
$$z_1{}^2 + z_2{}^2 + z_3{}^2 = z_1 z_2 + z_1 z_3 + z_2 z_3.$$

16.4 Answers to Odd-Numbered Exercises

(1) $-\frac{6}{25}$

(3) 7, 2

(5) 512, $\sqrt{3}$, 1024, $-\frac{\pi}{3}$

(7) $\frac{3}{2}$, $\frac{3\sqrt{3}}{2}$, -3, 0

(9) $2 - i$, $1 + 3i$

(11) 1, i, 0

(13) (a) 0, 1, 4
  (b) $2i$, $-2i$, $2i$
  (c) $2i$
  (d) $2i$
  (e) $\frac{1}{2}$, 2, $\cos 2t$, $\sin 2t$,
  (f) $3\cos 2t + \sin 2t$

Chapter 17

REAL AND COMPLEX INNER PRODUCT SPACES

17.1   Background

Topics: inner products in real and complex vector spaces, the *Schwarz inequality*, the *parallelogram law*, the *Pythagorean theorem*, the norm induced by an inner product, the metric induced by a norm, orthogonal (or perpendicular) vectors, the angle between two vectors, rowspace and columnspace of a matrix.

Definition 17.1.1. Let V be a (real or complex) vector space. A function that associates to each pair of vectors $\mathbf{x}$ and $\mathbf{y}$ in V a (real or complex) number $\langle \mathbf{x}, \mathbf{y} \rangle$ (often written $\mathbf{x} \cdot \mathbf{y}$) is an INNER PRODUCT (or a DOT PRODUCT) on V provided that the following four conditions are satisfied:

(a) If $\mathbf{x}$, $\mathbf{y}$, $\mathbf{z} \in V$, then

$$\langle \mathbf{x} + \mathbf{y}, \mathbf{z} \rangle = \langle \mathbf{x}, \mathbf{z} \rangle + \langle \mathbf{y}, \mathbf{z} \rangle.$$

(b) If $\mathbf{x}$, $\mathbf{y} \in V$ and $\alpha \in \mathbb{C}$ (or $\mathbb{R}$), then

$$\langle \alpha \mathbf{x}, \mathbf{y} \rangle = \alpha \langle \mathbf{x}, \mathbf{y} \rangle.$$

(c) If $\mathbf{x}$, $\mathbf{y} \in V$, then

$$\langle \mathbf{x}, \mathbf{y} \rangle = \overline{\langle \mathbf{y}, \mathbf{x} \rangle}.$$

(d) For every nonzero $\mathbf{x}$ in V we have $\langle \mathbf{x}, \mathbf{x} \rangle > 0$.

Conditions (a) and (b) show that an inner product is linear in its first variable. It is easy to see that an inner product is *conjugate linear* in its second variable. (A complex valued function f on V is CONJUGATE LINEAR if $f(\mathbf{x} + \mathbf{y}) = f(\mathbf{x}) + f(\mathbf{y})$ and $f(\alpha\mathbf{x}) = \overline{\alpha}f(\mathbf{x})$ for all $\mathbf{x}$, $\mathbf{y} \in V$ and $\alpha \in \mathbb{C}$.) When a mapping is linear in one variable and conjugate linear in the other, it is often called SESQUILINEAR (the prefix "sesqui-" means "one and a half"). Taken together conditions (a)–(d) say that the inner product is a *positive definite conjugate symmetric sesquilinear form*. When a vector space has been equipped with an inner product we define the NORM (or LENGTH) of a vector $\mathbf{x}$ by

$$\|\mathbf{x}\| := \sqrt{\langle \mathbf{x}, \mathbf{x} \rangle};$$

(Notice that this is the same definition we used in 3.1.2 for vectors in $\mathbb{R}^n$.)

Notation 17.1.2. There are many more or less standard notations for the inner product of two vectors $\mathbf{x}$ and $\mathbf{y}$. The two that we will use interchangeably in these exercises are $\mathbf{x} \cdot \mathbf{y}$ and $\langle \mathbf{x}, \mathbf{y} \rangle$.

Example 17.1.3. For vectors $x = (x_1, x_2, \ldots, x_n)$ and $y = (y_1, y_2, \ldots, y_n)$ belonging to $\mathbb{R}^n$ define

$$\langle x, y \rangle = \sum_{k=1}^{n} x_k y_k.$$

Then $\mathbb{R}^n$ is an inner product space.

Example 17.1.4. For vectors $x = (x_1, x_2, \ldots, x_n)$ and $y = (y_1, y_2, \ldots, y_n)$ belonging to $\mathbb{C}^n$ define

$$\langle x, y \rangle = \sum_{k=1}^{n} x_k \overline{y_k}.$$

Then $\mathbb{C}^n$ is an inner product space.

Example 17.1.5. Let l_2 be the set of all square summable sequences of complex numbers. (A sequence $x = (x_k)_{k=1}^{\infty}$ is SQUARE SUMMABLE if $\sum_{k=1}^{\infty} |x_k|^2 < \infty$.) (The vector space operations are defined pointwise.) For vectors $x = (x_1, x_2, \ldots)$ and $y = (y_1, y_2, \ldots)$ belonging to l_2 define

$$\langle x, y \rangle = \sum_{k=1}^{\infty} x_k \overline{y_k}.$$

Then l_2 is an inner product space. (It is important to recognize that in order for this definition to make sense, it must be verified that the infinite series actually converges.)

Example 17.1.6. For $a < b$ let $\mathcal{C}([a,b])$ be the family of all continuous complex valued functions on the interval $[a,b]$. For every $f, g \in \mathcal{C}([a,b])$ define

$$\langle f,g \rangle = \int_a^b f(x)\overline{g(x)}\,dx.$$

Then $\mathcal{C}([a,b])$ is an inner product space.

Definition 17.1.7. Let $\mathbf{x}$ and $\mathbf{y}$ be nonzero vectors in a real vector space V. Then $\angle(\mathbf{x}, \mathbf{y})$, the ANGLE between $\mathbf{x}$ and $\mathbf{y}$, is defined by

$$\angle(\mathbf{x}, \mathbf{y}) = \arccos \frac{\langle \mathbf{x}, \mathbf{y} \rangle}{\|\mathbf{x}\|\,\|\mathbf{y}\|}.$$

(Notice that this is the same definition as the one given in 3.1.3 for vectors in $\mathbb{R}^n$.)

Definition 17.1.8. Two vectors $\mathbf{x}$ and $\mathbf{y}$ in an inner product space V are ORTHOGONAL (or PERPENDICULAR) if $\langle \mathbf{x}, \mathbf{y} \rangle = 0$. In this case we write $\mathbf{x} \perp \mathbf{y}$. Similarly if M and N are nonempty subsets of V we write $M \perp N$ if every vector in M is orthogonal to every vector in N. When one of the sets consists of a single vector we write $x \perp N$ instead of $\{x\} \perp N$. When M is a nonempty subset of V we denote by $M^\perp$ the set of all vectors x such that $x \perp M$. This is the ORTHOGONAL COMPLEMENT of M.

Definition 17.1.9. A real valued function f on an inner product space V is UNIFORMLY CONTINUOUS if for every number $\epsilon > 0$ there exists a number $\delta > 0$ such that $|f(x) - f(y)| < \epsilon$ whenever $\|x - y\| < \delta$ in V.

The following result is one we have already seen for $\mathbb{R}^n$ (see 3.1.4).

Theorem 17.1.10 (Cauchy-Schwarz inequality). *If* $\mathbf{x}$ *and* $\mathbf{y}$ *are vectors in a (real or complex) inner product space, then*

$$|\langle \mathbf{x}, \mathbf{y} \rangle| \leq \|\mathbf{x}\|\,\|\mathbf{y}\|.$$

17.2   Exercises

(1) In $\mathbb{C}^3$ let $\mathbf{x} = (3 + 2i, 1 - i, i)$ and $\mathbf{y} = (1 - 2i, 1 - i, 4 + i)$. Then $\|\mathbf{x}\| = $ ____; $\|\mathbf{y}\| = $ ____; and $\langle \mathbf{x}, \mathbf{y} \rangle = $ _____.

(2) In $\mathbb{C}^2$ let $\mathbf{x} = (2 - 4i, 4i)$ and $\mathbf{y} = (2 + 4i, 4)$. Then $\|\mathbf{x}\| = $ ____; $\|\mathbf{y}\| = $ ____; and $\langle \mathbf{x}, \mathbf{y} \rangle = $ ____.

(3) In $\mathbb{C}^2$ let $\mathbf{x} = (3 - 2i, \sqrt{3}i)$ and $\mathbf{y} = (1 + i, 1 - i)$. Then $\|\mathbf{x}\| =$ _____;
$\|\mathbf{y}\| =$ _____; and $\langle \mathbf{x}, \mathbf{y} \rangle = 1 - \sqrt{a} + (\sqrt{a} - b)i$ where $a =$ ___ and
$b =$ ___.

(4) Make the family of 2×2 matrices of real numbers into an inner
product space by defining $\langle A, B \rangle := \text{tr}(A^t B)$ (see problem 8) for A,
$B \in \mathbf{M}_{2,2}(\mathbb{R})$. Let $U = \begin{bmatrix} 1 & 4 \\ -3 & 5 \end{bmatrix}$ and $V = \begin{bmatrix} \alpha^2 & \alpha - 1 \\ \alpha + 1 & -1 \end{bmatrix}$. Find all
values of α such that $U \perp V$ in the inner product space $\mathbf{M}_{2,2}(\mathbb{R})$.
Answer: $\alpha =$ _____ and _____.

(5) Let $\mathbf{w} = \left(\dfrac{i}{\sqrt{3}}, \dfrac{i}{\sqrt{3}}, \dfrac{i}{\sqrt{3}} \right)$, $\mathbf{x} = \left(-\dfrac{2i}{\sqrt{6}}, \dfrac{i}{\sqrt{6}}, \dfrac{i}{\sqrt{6}} \right)$, $\mathbf{y} =$
$\left(\dfrac{i}{\sqrt{6}}, \dfrac{i}{\sqrt{6}}, -\dfrac{2i}{\sqrt{6}} \right)$, and $\mathbf{z} = \left(0, -\dfrac{i}{\sqrt{2}}, \dfrac{i}{\sqrt{2}} \right)$.

(a) Which three of these vectors form an orthonormal basis for $\mathbb{C}^3$?
Answer: _____, _____, and _____.
(b) Write $(1, 0, 0)$ as a linear combination of the three basis vectors
you chose in part (a). (Use 0 as the coefficient of the vector that
does *not* belong to the basis.)
Answer: $(1, 0, 0) = -\dfrac{i}{\sqrt{a}} \mathbf{w} + \dfrac{2i}{\sqrt{2a}} \mathbf{x} + b\mathbf{y} + c\mathbf{z}$ where $a =$ _____,
$b =$ _____, and $c =$ _____.

(6) Find all real numbers α such that the angle between the vectors
$2\mathbf{i} + 2\mathbf{j} + (\alpha - 2)\mathbf{k}$ and $2\mathbf{i} + (\alpha - 2)\mathbf{j} + 2\mathbf{k}$ is $\frac{\pi}{3}$. Answer: $\alpha =$ _____
and _____.

(7) Let $\mathbf{f}(x) = x$ and $\mathbf{g}(x) = x^2$ for $0 \leq x \leq 1$. Then the cosine of the angle
between $\mathbf{f}$ and $\mathbf{g}$ in the inner product space $\mathcal{C}([0, 1])$ of all continuous
real valued functions on $[0, 1]$ is $\dfrac{a}{4}$ where $a =$ _____.

(8) Let $\mathbf{f}(x) = x$ and $\mathbf{g}(x) = \cos \pi x$ for $0 \leq x \leq 1$. In the inner product
space $\mathcal{C}([0, 1])$ of all continuous real valued functions on $[0, 1]$ the cosine
of the angle between $\mathbf{f}$ and $\mathbf{g}$ is $-\dfrac{a\sqrt{b}}{c^2}$ where $a =$ _____, $b =$ _____,
and $c =$ _____.

(9) Let $\mathbf{f}(x) = x^2$ and $\mathbf{g}(x) = 1 - cx$ where $0 \leq x \leq 1$ and c is a constant. If
$c =$ _____, then $\mathbf{f} \perp \mathbf{g}$ in the inner product space $\mathcal{C}([0, 1])$ of continuous
real valued function on $[0, 1]$.

(10) In $\mathbb{R}^4$ the subspace perpendicular to both $(1, 4, 4, 1)$ and $(2, 9, 8, 2)$ is the span of the vectors $(-4, a, b, a)$ and $(-b, a, a, b)$ where $a =$ _____ and $b =$ _____.

(11) A complex valued function $\mathbf{f}$ on the interval $[-\pi, \pi]$ is said to be SQUARE INTEGRABLE if $\int_{-\pi}^{\pi} |\mathbf{f}(x)|^2 \, dx < \infty$. Let $\mathcal{F}([-\pi, \pi])$ be the family of all square integrable complex valued functions on $[-\pi, \pi]$. This family is an inner product space under the usual pointwise operations of addition and scalar multiplication and the inner product defined by

$$\langle \mathbf{f}, \mathbf{g} \rangle = \frac{1}{2\pi} \int_{-\pi}^{\pi} \mathbf{f}(x) \overline{\mathbf{g}(x)} \, dx$$

for all $\mathbf{f}, \mathbf{g} \in \mathcal{F}([-\pi, \pi])$. Actually the preceding sentence is a lie: to be correct we should identify any two square integrable functions that differ only on a set of Lebesgue measure zero and work with the resulting equivalence classes. *Then* we have an inner product space. What we actually have here is a so-called *semi-inner product*. For the purposes of the current exercise, however, this correction turns out to be unimportant; ignore it.

For each integer n (positive, negative, or zero) define a function $\mathbf{e}_n$ by

$$\mathbf{e}_n(x) = e^{inx} \quad \text{for } -\pi \le x \le \pi.$$

(a) Then each $\mathbf{e}_n$ belongs to $\mathcal{F}([-\pi, \pi])$ and $\|\mathbf{e}_n\| =$ ___ for every integer n.

(b) If $m \ne n$, then $\langle \mathbf{e}_m, \mathbf{e}_n \rangle =$ ___.

Now let $\mathbf{f}(x) = \begin{cases} 0 & \text{if } -\pi \le x < 0, \\ 1 & \text{if } 0 \le x \le \pi. \end{cases}$

(c) Then $\langle \mathbf{f}, \mathbf{e}_0 \rangle =$ _____.

(d) If n is odd, then $\langle \mathbf{f}, \mathbf{e}_n \rangle =$ _____.

(e) If n is even but not zero, then $\langle \mathbf{f}, \mathbf{e}_n \rangle =$ _____.

(f) Write the sum of the middle eleven terms of the Fourier series for $\mathbf{f}$ in simplified form. *Hint:* Use problem 9 in this chapter.

Answer: $\sum_{n=-5}^{5} \langle \mathbf{f}, \mathbf{e}_n \rangle \mathbf{e}_n = a + b \sin x + c \sin 3x + d \sin 5x$ where

$a =$ _____, $b =$ _____, $c =$ _____, and $d =$ _____.

17.3 Problems

(1) Prove that if $\mathbf{x} \in V$ and $\langle \mathbf{x}, \mathbf{y} \rangle = 0$ for all $\mathbf{y} \in V$, then $\mathbf{x} = \mathbf{0}$.

(2) Let $S, T \colon V \to W$ be linear transformations between real inner product spaces. Prove that if $\langle S\mathbf{v}, \mathbf{w} \rangle = \langle T\mathbf{v}, \mathbf{w} \rangle$ for all $\mathbf{v} \in V$ and $\mathbf{w} \in W$, then $S = T$.

(3) Prove that if V is a complex inner product space and $T \in \mathcal{L}(V)$ satisfies $\langle T\mathbf{z}, \mathbf{z} \rangle = 0$ for all $\mathbf{z} \in V$, then $T = 0$. *Hint.* In the hypothesis replace $\mathbf{z}$ first by $\mathbf{x} + \mathbf{y}$ and then by $\mathbf{x} + i\mathbf{y}$.

(4) Show that the preceding result does not hold for real inner product spaces.

(5) Let V be a complex inner product space and $S, T \in \mathcal{L}(V)$. Prove that if $\langle S\mathbf{x}, \mathbf{x} \rangle = \langle T\mathbf{x}, \mathbf{x} \rangle$ for all $\mathbf{x} \in V$, then $S = T$.

(6) Prove the *Schwarz inequality* 17.1.10. *Hint.* Let $\alpha = \langle \mathbf{y}, \mathbf{y} \rangle$ and $\beta = -\langle \mathbf{x}, \mathbf{y} \rangle$ and expand $\|\alpha\mathbf{x} + \beta\mathbf{y}\|^2$. This hint leads to a slick, easy, and totally unenlightening proof. Perhaps you can find a more perspicuous one.

(7) [The *polarization identity*] If x and y are vectors in a complex inner product space, then

$$\langle x, y \rangle = \tfrac{1}{4}(\|x + y\|^2 - \|x - y\|^2 + i\,\|x + iy\|^2 - i\,\|x - iy\|^2).$$

What is the correct identity for a *real* inner product space?

(8) Let $\mathbf{M}_{2,2}(\mathbb{R})$ be the vector space of all 2×2 matrices of real numbers. Show that this space can be made into an inner product space by defining $\langle A, B \rangle := \mathrm{tr}(A^t B)$ for all $A, B \in \mathbf{M}_{2,2}(\mathbb{R})$.

(9) Prove that for every real number θ

$$e^{i\theta} = \cos\theta + i\sin\theta.$$

Derive from this that $\cos\theta = \tfrac{1}{2}\left(e^{i\theta} + e^{-i\theta}\right)$ and $\sin\theta = \tfrac{1}{2i}\left(e^{i\theta} - e^{-i\theta}\right)$.

(10) Let $\mathbf{x}$ and $\mathbf{y}$ be vectors in an inner product space. Prove that

$$\|\mathbf{x} + \mathbf{y}\|^2 + \|\mathbf{x} - \mathbf{y}\|^2 = 2\|\mathbf{x}\|^2 + 2\|\mathbf{y}\|^2.$$

Give a geometric interpretation of this result.

(11) Show that the norm function $\| \cdot \| \colon V \to \mathbb{R}$ on an inner product space is uniformly continuous.

17.4 Answers to Odd-Numbered Exercises

(1) $4, 2\sqrt{6}, 2 + 12i$

(3) $4, 2, 3, 5$

(5) (a) w, x, y
   (b) $3, 0, 0$

(7) $\sqrt{15}$

(9) $\dfrac{4}{3}$

(11) (a) 1
   (b) 0
   (c) $\dfrac{1}{2}$
   (d) $-\dfrac{i}{n\pi}$
   (e) 0
   (f) $\dfrac{1}{2}, \dfrac{2}{\pi}, \dfrac{2}{3\pi}, \dfrac{2}{5\pi}$

Chapter 18

ORTHONORMAL SETS
OF VECTORS

18.1 Background

Topics: orthonormal sets of vectors, orthonormal bases, orthogonal complements, orthogonal direct sums, *Gram-Schmidt orthonormalization*, the *QR-factorization* of a matrix.

Definition 18.1.1. A set B of vectors in an inner product space is ORTHONORMAL if $\mathbf{x} \perp \mathbf{y}$ whenever $\mathbf{x}$ and $\mathbf{y}$ are distinct vectors in B and $\|\mathbf{x}\| = 1$ for every $\mathbf{x} \in B$. The set B is a MAXIMAL ORTHONORMAL SET provided that it is orthonormal and the only orthonormal set which contains B is B itself.

Theorem 18.1.2. *Let $B = \{\mathbf{e}^1, \ldots, \mathbf{e}^n\}$ be an orthonormal set in an inner product space V. Then the following are equivalent.*

(a) *B is a maximal orthonormal set in V.*
(b) *If $\langle \mathbf{x}, \mathbf{e}^k \rangle = 0$ for $k = 1, \ldots, n$, then $\mathbf{x} = \mathbf{0}$.*
(c) *The span of B is all of V.*
(d) *If $\mathbf{x} \in V$, then $\mathbf{x} = \sum_{k=1}^{n} \langle \mathbf{x}, \mathbf{e}^k \rangle \mathbf{e}^k$.* (the Fourier series for $\mathbf{x}$.)
(e) *$\langle \mathbf{x}, \mathbf{y} \rangle = \sum_{k=1}^{n} \langle \mathbf{x}, \mathbf{e}^k \rangle \langle \mathbf{e}^k, \mathbf{y} \rangle$ for all $\mathbf{x}, \mathbf{y} \in V$.*
(f) *$\|\mathbf{x}\|^2 = \sum_{k=1}^{n} |\langle \mathbf{x}, \mathbf{e}^k \rangle|^2$ for every $\mathbf{x} \in V$.*
(g) *$\dim V = n$.*

Definition 18.1.3. An orthonormal set in an inner product space V that satisfies any (and hence all) of the conditions listed in the preceding theorem is called an ORTHONORMAL BASIS for V. In this case the scalars $\langle \mathbf{x}, \mathbf{e}^k \rangle$, that appear in Theorem 18.1.2 are the FOURIER COEFFICIENTS of $\mathbf{x}$ with respect to B.

Definition 18.1.4. An $n \times n$ matrix of real numbers is an ORTHOGONAL MATRIX if its column vectors are an orthonormal basis for $\mathbb{R}^n$. An $n \times n$ matrix of complex numbers is a UNITARY MATRIX if its column vectors are an orthonormal basis for $\mathbb{C}^n$.

Theorem 18.1.5 (*QR*-factorization). *If A is an $n \times n$ matrix of real numbers, then there exist an orthogonal matrix Q and an upper triangular matrix R such that $A = QR$. If A is an $n \times n$ matrix of complex numbers, then there exist a unitary matrix Q and an upper triangular matrix R such that $A = QR$.*

(For a proof of the preceding theorem see [8], pages 425–427.)

Definition 18.1.6. Let M and N be subspaces of an inner product space V. We say that the space V is the ORTHOGONAL DIRECT SUM of M and N if $M + N = V$ and $M \perp N$. In this case we write

$$V = M \oplus N.$$

Caution: Since the same notation is used for direct sums of vector spaces and orthogonal direct sums of inner product spaces, close attention should be paid to the context in which these concepts arise. For example, if M is the x-axis and N is the line $y = x$ in $\mathbb{R}^2$, is it true that $\mathbb{R}^2 = M \oplus N$? *Yes*, if $\mathbb{R}^2$ is regarded as a vector space. *No*, if it is regarded as an inner product space.

18.2   Exercises

(1) Use the *Gram-Schmidt procedure* to find an orthonormal basis for the subspace of $\mathbb{R}^4$ spanned by $\mathbf{w}^1 = (1, 0, 0, 0)$, $\mathbf{w}^2 = (1, 1, 1, 0)$, and $\mathbf{w}^3 = (1, 2, 0, 1)$. The basis consists of the vectors $\mathbf{e}^1 = (1, 0, 0, 0)$; $\mathbf{e}^2 = \dfrac{1}{a} (0, 1, 1, b)$; and $\mathbf{e}^3 = \dfrac{1}{c} (b, 1, -1, 1)$ where $a = \underline{\hspace{2em}}$, $b = \underline{\hspace{2em}}$, and $c = \underline{\hspace{2em}}$.

(2) Let $\mathcal{P}_4 = \mathcal{P}_4([0, 1])$ be the vector space of polynomials of degree strictly less than 4 with an inner product defined by

$$\langle \mathbf{p}, \mathbf{q} \rangle = \int_0^1 \mathbf{p}(x)\mathbf{q}(x)\, dx$$

for all $\mathbf{p}, \mathbf{q} \in \mathcal{P}_4$. Let $\mathbf{w}^0(x) = 1$, $\mathbf{w}^1(x) = x$, $\mathbf{w}^2(x) = x^2$, and $\mathbf{w}^3(x) = x^3$ for $0 \le x \le 1$. Use the *Gram-Schmidt process* to convert the ordered basis $\{\mathbf{w}^0, \mathbf{w}^1, \mathbf{w}^2, \mathbf{w}^3\}$ to an orthonormal basis $\{\mathbf{e}^0, \mathbf{e}^1, \mathbf{e}^2, \mathbf{e}^3\}$ for $\mathcal{P}_4$.

Answer: $\mathbf{e}^0(x) = $ _____

$\mathbf{e}^1(x) = \sqrt{a}(bx - 1)$ where $a = $ _____ and $b = $ _____

$\mathbf{e}^2(x) = \sqrt{a}(bx^2 - bx + 1)$ where $a = $ _____ and $b = $ _____

$\mathbf{e}^3(x) = \sqrt{a}(bx^3 - cx^2 + dx - 1)$ where $a = $ _____,

$b = $ _____, $c = $ _____ , and $d = $ _____.

(3) Find the QR factorization of $A = \begin{bmatrix} 3 & 0 \\ 4 & 5 \end{bmatrix}$.

Answer: $A = QR = \dfrac{1}{a}\begin{bmatrix} b & -c \\ c & b \end{bmatrix}\begin{bmatrix} a & c \\ 0 & b \end{bmatrix}$ where $a = $ _____, $b = $ _____, and $c = $ _____.

(4) Let $A = \begin{bmatrix} 0 & 0 & 1 \\ 0 & 1 & 1 \\ 1 & 1 & 1 \end{bmatrix}$. The QR-factorization of A is $A = QR$

where $Q = \begin{bmatrix} & & \\ & & \\ & & \end{bmatrix}$ and $R = \begin{bmatrix} & & \\ & & \\ & & \end{bmatrix}$.

(5) Let $A = \begin{bmatrix} 1 & 4 & -2 \\ 1 & 3 & -1 \\ 1 & 2 & -1 \end{bmatrix}$. The QR-factorization of A is $A = QR$

where $Q = \begin{bmatrix} a & b & -ab \\ a & 0 & 2ab \\ a & -b & -ab \end{bmatrix}$ and $R = \begin{bmatrix} 3a & 9a & -4a \\ 0 & 2b & -b \\ 0 & 0 & ab \end{bmatrix}$ where $a = $ _____

and $b = $ _____.

18.3 Problems

(1) Let $\{e^1, e^2, \ldots, e^n\}$ be a finite orthonormal subset of an inner product space V and $\mathbf{x} \in V$. Show that

$$\sum_{k=1}^{n} |\langle \mathbf{x}, e^k \rangle|^2 \leq \|\mathbf{x}\|^2.$$

Hint. Multiply out $\left\langle \mathbf{x} - \sum_{k=1}^{n} \langle \mathbf{x}, e^k \rangle e^k, \mathbf{x} - \sum_{k=1}^{n} \langle \mathbf{x}, e^k \rangle e^k \right\rangle$.

(2) Let M be a subspace of an inner product space V.
   (a) Show that $M \subseteq M^{\perp\perp}$.
   (b) Prove that equality need not hold in (a).
   (c) Show that if V is finite dimensional, then $M = M^{\perp\perp}$.

(3) Let M and N be subspaces of an inner product space. Prove that

$$(M + N)^{\perp} = M^{\perp} \cap N^{\perp}.$$

(4) Let M be a subspace of a finite dimensional inner product space V. Prove that $V = M \oplus M^{\perp}$.

(5) Give an example to show that the conclusion of the preceding problem need not hold in an infinite dimensional space.

(6) Prove that if an inner product space V is the orthogonal direct sum $M \oplus N$ of two subspaces M and N, then $N = M^{\perp}$.

(7) Prove that if f is a linear functional on a finite dimensional inner product space V, then there exists a unique vector $\mathbf{a} \in V$ such that

$$f(\mathbf{x}) = \langle \mathbf{x}, \mathbf{a} \rangle$$

for every $\mathbf{x} \in V$.

(8) In beginning calculus, you found (by making use of the p-test) that the series $\sum_{k=1}^{\infty} \frac{1}{k^2}$ converges. But you were not given a means of discovering what the series converges to. Now you have enough machinery to accomplish this.

   We denote by $L_2[0, 2\pi]$ the vector space of all complex valued functions $\mathbf{f}$ defined on the interval $[0, 2\pi]$ such that

$$\int_0^{2\pi} |\mathbf{f}(t)|^2 \, dt < \infty.$$

(As in Exercise 11 of Chapter 17 this isn't quite correct: the members of L_2 are technically equivalence classes of functions. For the purposes of this problem, use the preceding not-quite-right definition.)

On the space $L_2[0, 2\pi]$ we define the following inner product

$$\langle \mathbf{f}, \mathbf{g} \rangle = \frac{1}{2\pi} \int_0^{2\pi} \mathbf{f}(t)\overline{\mathbf{g}(t)}\, dt.$$

For each integer n (positive, negative, or zero) define the function $\mathbf{e}^n$ on $[0, 2\pi]$ by

$$\mathbf{e}^n(x) = e^{inx}$$

for $0 \leq x \leq 2\pi$.

(a) Show that $\{\mathbf{e}^n : n \text{ is an integer}\}$ is an orthonormal set in $L_2[0, 2\pi]$.

In part (b) you may use without proof the following fact: for every function $\mathbf{f}$ in the inner product space $L_2[0, 2\pi]$

$$\|\mathbf{f}\|^2 = \sum_{k=-\infty}^{\infty} |\langle \mathbf{f}, \mathbf{e}^k \rangle|^2. \tag{$*$}$$

That is, in $L_2[0, 2\pi]$ the square of the length of a vector is the sum of the squares of its Fourier coefficients with respect to the orthonormal family given in part (a). This is the infinite dimensional version of *Parseval's formula*.

(b) Find the sum of the infinite series $\displaystyle\sum_{k=1}^{\infty} \frac{1}{k^2}$. *Hint.* Apply $(*)$ to the function $\mathbf{f}(x) = x$.

18.4 Answers to Odd-Numbered Exercises

(1) $\sqrt{2}$, 0, $\sqrt{3}$

(3) 5, 3, 4

(5) $\dfrac{1}{\sqrt{3}}, \dfrac{1}{\sqrt{2}}$

Chapter 19

QUADRATIC FORMS

19.1 Background

Topics: quadratic forms, quadric surfaces, positive (and negative) definite, positive (and negative) semidefinite, indefinite.

Definition 19.1.1. A symmetric matrix A on a real inner product space V is POSITIVE DEFINITE if $\langle A\mathbf{x}, \mathbf{x} \rangle > 0$ for every $\mathbf{x} \neq \mathbf{0}$ in V. It is NEGATIVE DEFINITE if $\langle A\mathbf{x}, \mathbf{x} \rangle < 0$ for every $\mathbf{x} \neq \mathbf{0}$ in V. It is POSITIVE SEMIDEFINITE if $\langle A\mathbf{x}, \mathbf{x} \rangle \geq 0$ for every $\mathbf{x} \neq \mathbf{0}$ in V. It is NEGATIVE SEMIDEFINITE if $\langle A\mathbf{x}, \mathbf{x} \rangle \leq 0$ for every $\mathbf{x} \neq \mathbf{0}$ in V. It is INDEFINITE if there are vectors $\mathbf{x}$ and $\mathbf{y}$ in V such that $\langle A\mathbf{x}, \mathbf{x} \rangle > 0$ and $\langle A\mathbf{y}, \mathbf{y} \rangle < 0$. Of course an operator on a finite dimensional vector space is positive definite, negative definite, or indefinite if its matrix representation is positive definite, *etc.*

The following useful result (and its proof) can be found on page 250 of [9].

Theorem 19.1.2. *Let A be a symmetric $n \times n$ matrix of real numbers. Then the following conditions are equivalent:*

(a) *A is positive definite;*
(b) *$\mathbf{x}^t A \mathbf{x} > 0$ for every $\mathbf{x} \neq \mathbf{0}$ in $\mathbb{R}^n$;*
(c) *every eigenvalue λ of A is strictly positive;*
(d) *every leading principal submatrix A_k $(k = 1, \ldots, n)$ has strictly positive determinant; and*
(e) *when A has been put in echelon form (without row exchanges) the pivots are all strictly positive.*

In the preceding, the *leading principal submatrix* A_k is the $k \times k$ matrix that appears in the upper left corner of A.

19.2 Exercises

(1) Suppose that A is a 3×3 matrix such that $\langle A\mathbf{x}, \mathbf{x} \rangle = x_1{}^2 + 5x_2{}^2 - 3x_3{}^2 + 6x_1x_2 - 4x_1x_3 + 8x_2x_3$ for all $\mathbf{x} \in \mathbb{R}^3$. Then $A = \begin{bmatrix} a & b & c \\ b & d & e \\ c & e & f \end{bmatrix}$
where $a = $ ___, $b = $ ___, $c = $ ___, $d = $ ___, $e = $ ___, and $f = $ ___.

(2) A curve C is given by the equation $2x^2 - 72xy + 23y^2 = 50$. What kind of curve is C?
Answer: It is $a(n)$ _____.

(3) The equation $5x^2 + 8xy + 5y^2 = 1$ describes an ellipse. The principal axes of the ellipse lie along the lines $y = $ ___ and $y = $ ___.

(4) The graph of the equation $13x^2 - 8xy + 7y^2 = 45$ is an ellipse. The length of its semimajor axis is ___ and the length of its semiminor axis is ___.

(5) Consider the equation $2x^2 + 2y^2 - z^2 - 2xy + 4xz + 4yz = 3$.

    (a) The graph of the equation is what type of quadric surface?
        Answer: _____.

    (b) In standard form the equation for this surface is
        $u^2 + v^2 + $ ___ $w^2 = $ ___.

    (c) Find three orthonormal vectors with the property that in the coordinate system they generate, the equation of the surface is in standard form.
        Answer: $\dfrac{1}{\sqrt{6}}(1,$ ___, ___$)$, $\dfrac{1}{\sqrt{2}}($ ___, ___, $0)$, and $\dfrac{1}{\sqrt{3}}(1,$ ___, ___$)$.

(6) Determine for each of the following matrices whether it is positive definite, positive semidefinite, negative definite, negative semidefinite, or indefinite.

    (a) The matrix $\begin{bmatrix} 2 & -1 & -1 \\ -1 & 2 & -1 \\ -1 & -1 & 2 \end{bmatrix}$ is _____.

(b) The matrix $\begin{bmatrix} -2 & 1 & 1 \\ 1 & -2 & -1 \\ 1 & -1 & -2 \end{bmatrix}$ is _____.

(7) Determine for each of the following matrices whether it is positive definite, positive semidefinite, negative definite, negative semidefinite, or indefinite.

(a) The matrix $\begin{bmatrix} 1 & 2 & 3 \\ 2 & 5 & 4 \\ 3 & 4 & 9 \end{bmatrix}$ is _____.

(b) The matrix $\begin{bmatrix} 1 & 2 & 0 & 0 \\ 2 & 6 & -2 & 0 \\ 0 & -2 & 5 & -2 \\ 0 & 0 & -2 & 3 \end{bmatrix}$ is _____.

(c) The matrix $\begin{bmatrix} 0 & 1 & 2 \\ 1 & 0 & 1 \\ 2 & 1 & 0 \end{bmatrix}^2$ is _____.

(8) Let $B = \begin{bmatrix} 2 & 2 & 4 \\ 2 & b & 8 \\ 4 & 8 & 7 \end{bmatrix}$. For what range of values of b is B positive definite?

Answer: _____.

(9) Let $A = \begin{bmatrix} a & 1 & 1 \\ 1 & a & 1 \\ 1 & 1 & a \end{bmatrix}$. For what range of values of a is A positive definite?

Answer: _____.

19.3   Problem

(1) You are given a quadratic form $q(x, y, z) = ax^2 + by^2 + cz^2 + 2dxy + 2exz + 2fyz$. Explain in detail how to determine whether the associated level surface $q(x, y, z) = c$ encloses a region of finite volume in $\mathbb{R}^3$ and, if it does, how to find that volume. Justify carefully all claims you make. Among other things, explain how to use the *change of variables theorem* for multiple integrals to express the volume of an ellipsoid in terms of the lengths of the principal axes of the ellipsoid.

Apply the method you have developed to the equation

$$11x^2 + 4y^2 + 11z^2 + 4xy - 10xz + 4yz = 8.$$

19.4   Answers to Odd-Numbered Exercises

(1) 1, 3, −2, 5, 4, −3

(3) −x, x

(5) (a) hyperboloid of one sheet
   (b) −1, 1
   (c) 1, −2, 1, −1, 1, 1

(7) (a) indefinite
   (b) positive definite
   (c) positive definite

(9) $a > 1$

Chapter 20

OPTIMIZATION

20.1 Background

Topics: critical (stationary) points of a function of several variables; local (relative) maxima and minima; global (absolute) maxima and minima.

Definition 20.1.1. Let $f\colon \mathbb{R}^n \to \mathbb{R}$ be a smooth scalar field (that is, a real valued function on $\mathbb{R}^n$ with derivatives of all orders) and $\mathbf{p} \in \mathbb{R}^n$. The HESSIAN MATRIX (or SECOND DERIVATIVE MATRIX) of f at $\mathbf{p}$, denoted by $H_f(\mathbf{p})$, is the symmetric $n \times n$ matrix

$$H_f(\mathbf{p}) = \left[\frac{\partial^2 f}{\partial x_i \partial x_j}(\mathbf{p}) \right]_{i=1\,j=1}^{n\ \ n} = \left[f_{ij}(\mathbf{p}) \right].$$

Theorem 20.1.2 (Second Derivative Test). *Let $\mathbf{p}$ be a critical point of a smooth scalar field f (that is, a point where the gradient of f is zero). If the Hessian matrix H_f is positive definite at $\mathbf{p}$, then f has a local minimum there. If H_f is negative definite at $\mathbf{p}$, then f has a local maximum there. If H_f is indefinite at $\mathbf{p}$, then f has a saddle point there.*

20.2 Exercises

(1) Notice that the function f defined by $f(x, y) = (x^2 - 2x) \cos y$ has a critical point (stationary point) at the point $(1, \pi)$. The eigenvalues of the Hessian matrix of f are _____ and _____ ; so we conclude that the point $(1, \pi)$ is a _____ (local minimum, local maximum, saddle point).

(2) Use matrix methods to classify the critical point of the function
$$f(x, y) = 2x^2 + 2xy + 2x + y^4 - 4y^3 + 7y^2 - 4y + 5$$
as a local maximum, local minimum, or saddle point.

(a) The only critical point is located at (_____, _____).

(b) It is a _____.

(3) Use matrix methods to classify the critical point of the function
$$f(x, y, z) = \frac{1}{2}x^4 - xy + y^2 - xz + z^2 - x + 3$$
as a local maximum, local minimum, or saddle point.

(a) The only critical point is located at (_____, _____, _____).

(b) It is a _____.

(4) Notice that the function f defined by $f(x, y) = -1 + 4(e^x - x) - 5x \sin y + 6y^2$ has a critical point (stationary point) at the origin. Since the eigenvalues of the Hessian matrix of f are _____ (both positive, both negative, of different signs) we conclude that the origin is a _____ (local minimum, local maximum, saddle point).

(5) Use matrix methods to classify each critical point of the function
$$f(x, y) = y^3 - \frac{4}{3}x^3 - 2y^2 + 2x^2 + y - 7$$
as a local maximum, local minimum, or saddle point.

Answer: $(0, \frac{1}{3})$ is a _____.

$(0, \underline{\ \ })$ is a _____.

$(\underline{\ \ }, \frac{1}{3})$ is a _____.

$(\underline{\ \ }, \underline{\ \ })$ is a _____.

(6) Use matrix methods to classify each critical point of the function
$$f(x, y, z) = x^2y - 4x - y \sin z \qquad \text{for } 0 < z < \pi$$
as a local maximum, local minimum, or saddle point.

Answer: The critical points are $(-1, \underline{\ \ \ \ }, \underline{\ \ \ \ })$, which is a _____; and $(1, \underline{\ \ \ \ }, \underline{\ \ \ \ })$, which is a _____ (local minimum, local maximum, saddle point).

(7) The function f defined by $f(x, y) = x^2y^2 - 2x - 2y$ has a stationary point at (_____, _____). At this stationary point f has a _____ (local minimum, local maximum, saddle point).

20.3 Problems

(1) Use matrix methods to classify each critical point of the function
$$f(x, y, z) = x^3y + z^2 - 3x - y + 4z + 5$$
as a local maximum, local minimum, or saddle point. Justify your conclusions carefully.

(2) Let $f(x, y, z) = x^2y - ye^z + 2x + z$. The only critical point of f is located at $(-1, 1, 0)$. Use the second derivative test to classify this point as a local maximum, local minimum, or saddle point. State the reasons for your conclusion clearly.

(3) Notice that the function f defined by $f(x, y, z) = x^2y + 2xy + y - ye^{z-1} + 2x + z + 7$ has a critical point (stationary point) at $(-2, 1, 1)$. Use the second derivative test to classify this point as a local maximum, local minimum, or saddle point. State the reasons for your conclusion clearly.

(4) Explain in detail how to use matrix methods to classify each critical point of the function
$$f(x, y) = -\frac{1}{2}xy + \frac{2}{x} - \frac{1}{y}$$
as a local maximum, local minimum, or saddle point. Carry out the computations you describe.

20.4 Answers to Odd-Numbered Exercises

(1) $-2, -1$, local maximum

(3) (a) $1, \dfrac{1}{2}, \dfrac{1}{2}$
    (b) local minimum

(5) saddle point, 1, local minimum, 1, local maximum, 1, 1, saddle point

(7) 1, 1, saddle point

Part 6

ADJOINT OPERATORS

Chapter 21

ADJOINTS AND TRANSPOSES

21.1 Background

Topics: adjoint of an operator, transpose of an operator, conjugate transpose.

Definition 21.1.1. Let $T\colon V \to W$ be a linear transformation between real inner product spaces. If there exists a linear map $T^t\colon W \to V$ that satisfies

$$\langle T\mathbf{v}, \mathbf{w} \rangle = \langle \mathbf{v}, T^t\mathbf{w} \rangle$$

for all $\mathbf{v} \in V$ and $\mathbf{w} \in W$, then T^t is the TRANSPOSE of T.

In connection with the definition above see Problem 1.

Theorem 21.1.2. *Let $T\colon \mathbb{R}^n \to \mathbb{R}^m$ be a linear transformation. Then the transpose linear transformation T^t exists. Furthermore, the matrix representation $[T^t]$ of this transformation is the transpose of the matrix representation of T.*

Definition 21.1.3. Let V be a real inner product space and T be an operator on V whose transpose exists. If $T = T^t$, then T is SYMMETRIC. If T commutes with its transpose (that is, if $TT^t = T^tT$) it is NORMAL.

Definition 21.1.4. If $[a_{ij}]$ is an $m \times n$ matrix, its CONJUGATE TRANSPOSE is the $n \times m$ matrix $[\overline{a_{ji}}]$.

Definition 21.1.5. Let $T: V \to W$ be a linear transformation between complex inner product spaces. If there exists a linear map $T^*: W \to V$ that satisfies

$$\langle T\mathbf{v}, \mathbf{w} \rangle = \langle \mathbf{v}, T^*\mathbf{w} \rangle$$

for all $\mathbf{v} \in V$ and $\mathbf{w} \in W$, then T^* is the ADJOINT (or CONJUGATE TRANSPOSE, or HERMITIAN CONJUGATE) of T. (In many places, T^* is denoted by T^H or by $T^\dagger$.)

Theorem 21.1.6. *Let $T: \mathbb{C}^n \to \mathbb{C}^m$ be a linear transformation. Then the adjoint linear transformation T^* exists. Furthermore, the matrix representation $[T^*]$ of this transformation is the conjugate transpose of the matrix representation of T.*

Definition 21.1.7. Let V be a complex inner product space and T be an operator on V whose adjoint exists. If $T = T^*$, then T is SELF-ADJOINT (or HERMITIAN). If T commutes with its adjoint (that is, if $TT^* = T^*T$) it is NORMAL. A matrix is NORMAL if it is the representation of a normal operator.

Definition 21.1.8. Let V and W be inner product spaces. We make the vector space $V \oplus W$ into an inner product space as follows. For $\mathbf{v}_1, \mathbf{v}_2 \in V$ and $\mathbf{w}_1, \mathbf{w}_2 \in W$ let

$$\langle (\mathbf{v}_1, \mathbf{w}_1), (\mathbf{v}_2, \mathbf{w}_2) \rangle := \langle \mathbf{v}_1, \mathbf{v}_2 \rangle + \langle \mathbf{w}_1, \mathbf{w}_2 \rangle.$$

(It is an easy exercise to verify that this is indeed an inner product on $V \oplus W$.)

21.2 Exercises

(1) Let $\mathcal{C}([0,1], \mathbb{C})$ be the family of all continuous complex valued functions on the interval $[0,1]$. The usual inner product on this space is given by

$$\langle \mathbf{f}, \mathbf{g} \rangle = \int_0^1 \mathbf{f}(x)\overline{\mathbf{g}(x)}\, dx.$$

Let ϕ be a fixed continuous complex valued function on $[0,1]$. Define the operator M_ϕ on the complex inner product space $\mathcal{C}([0,1], \mathbb{C})$ by $M_\phi(\mathbf{f}) = \phi\mathbf{f}$. Then

$$M_\phi{}^* = \underline{\hspace{2cm}}.$$

(2) Let $A = \begin{bmatrix} 3-i & 2+2i \\ 1 & 3i \\ \overline{1-i} & \end{bmatrix}$. Find Hermitian (that is, self-adjoint) matrices B and C such that $A = B + iC$. *Hint.* Consider $A \pm A^*$.

Answer: $B = \dfrac{1}{a} \begin{bmatrix} 4c & b+ci \\ b-ci & d \end{bmatrix}$ and $C = \dfrac{1}{a} \begin{bmatrix} -a & b-ci \\ b+ci & 4c \end{bmatrix}$, where

$a = $ _____, $b = $ _____, $c = $ _____, and $d = $ _____.

(3) Let $\mathcal{P}_3$ be the space of polynomial functions of degree strictly less than 3 defined on the interval $[0,1]$. Define the inner product of two polynomials $p, q \in \mathcal{P}_3$ by $\langle p, q \rangle = \int_0^1 p(t)q(t)\, dt$. Then the matrix representation of the transpose of the differentiation operator D on the space $\mathcal{P}_3$ (with

respect to its usual basis $\{1, t, t^2\}$) is $\begin{bmatrix} & & \\ & & \\ & & \end{bmatrix}$. *Hint.*

The answer is **not** the transpose of the matrix representation of D.

(4) Let V be a complex inner product space. Define an operator $T \colon V \oplus V \to V \oplus V$ by

$$T(\mathbf{x}, \mathbf{y}) = (\mathbf{y}, -\mathbf{x}).$$

Then $T^*(\mathbf{u}, \mathbf{v}) = ($ _____, _____ $)$.

21.3 Problems

(1) Let $T \colon V \to W$ be a linear map between real inner product spaces. If $S \colon W \to V$ is a function that satisfies

$$\langle T\mathbf{v}, \mathbf{w} \rangle = \langle \mathbf{v}, S\mathbf{w} \rangle$$

for all $\mathbf{v} \in V$ and all $\mathbf{w} \in W$, then S is linear (and is therefore the transpose of T).

(2) Prove Theorem 21.1.2. Show that, in fact, every linear map between finite dimensional real inner product spaces has a transpose.

(3) Let T be a self-adjoint operator on a complex inner product space V. Prove that $\langle T\mathbf{x}, \mathbf{x} \rangle$ is real for every $\mathbf{x} \in V$.

(4) Let T be an operator on a complex inner product space whose adjoint T^* exists. Prove that $T^*T = 0$ if and only if $T = \mathbf{0}$.

(5) Let V be a complex inner product space and let ϕ be defined on the set $\mathfrak{A}(V)$ of operators on V whose adjoint exists by

$$\phi(T) = T^*.$$

Show that if $S,\, T \in \mathfrak{A}(V)$ and $\alpha \in \mathbb{C}$, then $(S+T)^* = S^* + T^*$ and $(\alpha T)^* = \overline{\alpha}\, T^*$. *Hint.* Use Problem 5 in Chapter 17.
Note: Similarly, if V is a real inner product space, $\mathfrak{A}(V)$ is the set of operators whose transpose exists, and $\phi(T) := T^t$, then ϕ is linear.

(6) Let T be a linear operator on a complex inner product space V. Show if T has an adjoint, then so does T^* and $T^{**} = T$. *Hint:* Use Problem 5 in Chapter 17. (Here T^{**} means $\left(T^*\right)^*$.)
Note: The real inner product space version of this result says that if T is an operator on a real inner product space whose transpose exists, then the transpose of T^t exists and $T^{tt} = T$.

(7) Let S and T be operators on a complex inner product space V. Show that if S and T have adjoints, then so does ST and $(ST)^* = T^*S^*$. *Hint.* Use Problem 5 in Chapter 17.
Note: The real inner product space version of this says that if S and T are operators on a real inner product space and if S and T both have transposes, then so does ST and $(ST)^t = T^tS^t$.

(8) Let $A\colon V \to V$ be an operator on a real inner product space. Suppose that A^t exists and that it commutes with A (that is, suppose $AA^t = A^tA$). Show that $\ker A = \ker A^t$.

(9) Let A and B be Hermitian operators on a complex inner product space. Prove that AB is Hermitian if and only if $AB = BA$.

(10) Show that if $T\colon V \to W$ is an invertible linear map between complex inner product spaces and both T and T^{-1} have adjoints, then T^* is invertible and $(T^*)^{-1} = (T^{-1})^*$.
Note: The real inner product space version of this says that if $T\colon V \to W$ is an invertible linear map between real inner product spaces and both T and T^{-1} have transposes, then T^t is invertible and $(T^t)^{-1} = (T^{-1})^t$.

(11) Every eigenvalue of a self-adjoint operator on a complex inner product space is real. *Hint.* Let $\mathbf{x}$ be an eigenvector associated with an eigenvalue λ of an operator A. Consider $\lambda \|\mathbf{x}\|^2$.

(12) Let A be a self-adjoint operator on a complex inner product space. Prove that eigenvectors associated with distinct eigenvalues of A are orthogonal. *Hint.* Use problem 11. Let x and y be eigenvectors associated with distinct eigenvalues λ and μ of A. Start your proof by showing that $\lambda\langle x, y \rangle = \mu\langle x, y \rangle$.

21.4 Answers to Odd-Numbered Exercises

(1) $M_{\bar{\phi}}$

(3) $\begin{bmatrix} -6 & 2 & 3 \\ 12 & -24 & -26 \\ 0 & 30 & 30 \end{bmatrix}$

Chapter 22

THE FOUR FUNDAMENTAL SUBSPACES

22.1 Background

Topics: column space; row space; nullspace; left nullspace, lead variables and free variables in a matrix, rank, row rank and column rank of a matrix.

Definition 22.1.1. A linear transformation $T \colon \mathbb{R}^n \to \mathbb{R}^m$ (and its associated standard matrix $[T]$) have *four* fundamental subspaces: the kernels and ranges of T and T^t. Over the years a rather elaborate terminology has grown up around these basic notions.

The NULLSPACE of the matrix $[T]$ is the kernel of the linear map T.

The LEFT NULLSPACE of the matrix $[T]$ is the kernel of the linear map T^t.

The COLUMN SPACE of $[T]$ is the subspace of $\mathbb{R}^m$ spanned by the column vectors of the matrix $[T]$. This is just the range of the linear map T.

And finally, the ROW SPACE of $[T]$ is the subspace of $\mathbb{R}^n$ spanned by the row vectors of the matrix $[T]$. This is just the range of the linear map T^t.

For a linear transformation $T \colon \mathbb{C}^n \to \mathbb{C}^m$ the terminology is the same. **EXCEPT:** in the preceding five paragraphs each appearance of "T^t" must be replaced by a "T^*" (and, of course, $\mathbb{R}^n$ by $\mathbb{C}^n$ and $\mathbb{R}^m$ by $\mathbb{C}^m$).

Definition 22.1.2. The ROW RANK of a matrix is the dimension of its row space and the COLUMN RANK of a matrix is the dimension of its column space.

Proposition 22.1.3. *The rank of a matrix A is the dimension of the largest square submatrix of A with nonzero determinant.*

161

Two useful facts that you may wish to keep in mind are:

(i) row equivalent matrices have the same row space (for a proof see [6], page 56); and

(ii) the row rank of a matrix is the same as its column rank (for a proof see [6], page 72).

Note that according to the second assertion the rank of a linear map T, the row rank of its matrix representation $[T]$, and the column rank of $[T]$ are all equal.

Theorem 22.1.4 (Fundamental Theorem of Linear Algebra). *If T is an operator on a finite dimensional complex inner product space, then*

$$\ker T = (\operatorname{ran} T^*)^{\perp}.$$

Corollary 22.1.5. *If T is an operator on a finite dimensional complex inner product space, then*

$$\ker T^* = (\operatorname{ran} T)^{\perp}.$$

Corollary 22.1.6. *If T is an operator on a finite dimensional complex inner product space, then*

$$\operatorname{ran} T = (\ker T^*)^{\perp}.$$

Corollary 22.1.7. *If T is an operator on a finite dimensional complex inner product space, then*

$$\operatorname{ran} T^* = (\ker T)^{\perp}.$$

Note: With the obvious substitutions of T^t for T^*, the preceding theorem and its three corollaries remain true for finite dimensional real inner product spaces.

22.2 Exercises

(1) Let $T\colon \mathbb{C}^3 \to \mathbb{C}^3$ be the operator whose matrix representation is

$$[T] = \begin{bmatrix} 1 & i & -1 \\ 1+i & 3-i & -2 \\ 1-2i & -6+5i & 1 \end{bmatrix}.$$

(a) The kernel of T (the nullspace of $[T]$) is the span of

$$\{(\underline{\hspace{1cm}}, \underline{\hspace{1cm}}, 10)\}.$$

(b) The range of T (the column space of $[T]$) is the span of

$$\{(1, 0, \underline{\hspace{0.5cm}}), (0, 1, \underline{\hspace{0.5cm}})\}.$$

(c) The kernel of T^* (the left nullspace of T^*) is the span of

$$\{(3, \underline{\hspace{0.5cm}}, \underline{\hspace{0.5cm}})\}.$$

(d) The range of T^* (the row space of $[T]$) is the span of

$$\{(10, 0, \underline{\hspace{1cm}}), (0, 10, \underline{\hspace{1cm}})\}.$$

(2) Find a basis for each of the four fundamental subspaces associated with the matrix

$$A = \begin{bmatrix} 1 & 2 & 0 & 1 \\ 0 & 1 & 1 & 0 \\ 1 & 2 & 0 & 1 \end{bmatrix}.$$

(a) The column space of A is the plane in $\mathbb{R}^3$ whose equation is
_____. It is the span of $\{(1, 0, \underline{\hspace{0.5cm}}), (0, 1, \underline{\hspace{0.5cm}})\}$.

(b) The nullspace of A is the span of $\{(\underline{\hspace{0.5cm}}, -1, \underline{\hspace{0.5cm}}, 0), (\underline{\hspace{0.5cm}}, 0, 0, 1)\}$.

(c) The row space of A is the span of $\{(1, 0, \underline{\hspace{0.5cm}}, 1), (0, \underline{\hspace{0.5cm}}, 1, 0)\}$.

(d) The left nullspace of A is the line in $\mathbb{R}^3$ whose equations are
$\underline{\hspace{1cm}} = \underline{\hspace{0.5cm}} = 0$. It is the span of $\{(\underline{\hspace{0.5cm}}, \underline{\hspace{0.5cm}}, 1)\}$.

(3) Let $A = \begin{bmatrix} 1 & 2 & 0 & 2 & -1 & 1 \\ 3 & 6 & 1 & 1 & -2 & 1 \\ 5 & 10 & 1 & 5 & -4 & 3 \end{bmatrix}.$

(a) Find a basis for the column space of A.

Answer:

$$\{(1, 0, \underline{\hspace{0.5cm}}), (0, \underline{\hspace{0.5cm}}, 1)\}.$$

(b) The column space of A is a plane in $\mathbb{R}^3$. What is its equation?

Answer: _____.

(c) The dimension of the row space of A is _____.

(d) Fill in the missing coordinates of the following vector so that it lies in the row space of A.

$$(4, \underline{\hspace{0.6cm}}, 6, \underline{\hspace{0.9cm}}, \underline{\hspace{0.6cm}}, \underline{\hspace{0.6cm}}).$$

(e) The dimension of the nullspace of A is \underline{\hspace{0.6cm}}.

(f) Fill in the missing coordinates of the following vector so that it lies in the nullspace of A.

$$(\underline{\hspace{0.8cm}}, 1, \underline{\hspace{0.8cm}}, 1, 1, 1).$$

(4) Let $T\colon \mathbb{R}^5 \to \mathbb{R}^3$ be defined by

$$T(v, w, x, y, z) = (v - x + z, v + w - y, w + x - y - z).$$

(a) Find the matrix representation of T.

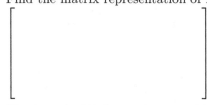

(b) The kernel of T (the nullspace of $[T]$) is the span of

$$\{(\underline{\hspace{0.4cm}}, \underline{\hspace{0.4cm}}, 1, 0, 0), (0, 1, 0, 1, 0), (-1, 1, 0, \underline{\hspace{0.4cm}}, 1)\}.$$

(c) The range of T (the column space of $[T]$) is the span of

$$\{(1, \underline{\hspace{0.4cm}}, -1), (0, 1, \underline{\hspace{0.4cm}})\}.$$

Geometrically this is a \underline{\hspace{4cm}}.

(d) The range of T^t (the row space of $[T]$) is the span of

$$\{(1, 0, \underline{\hspace{0.4cm}}, 0, 1), (\underline{\hspace{0.4cm}}, \underline{\hspace{0.4cm}}, \underline{\hspace{0.4cm}}, -1, -1)\}.$$

(e) The kernel of T^t (the left nullspace of $[T]$) is the span of

$$\{(1, \underline{\hspace{0.4cm}}, \underline{\hspace{0.4cm}})\}.$$

Geometrically this is a \underline{\hspace{4cm}}.

(5) Let A be the matrix $\begin{bmatrix} 1 & 0 & 2 & 0 & -1 \\ 1 & 2 & 4 & -2 & -1 \\ 0 & 1 & 1 & -1 & 0 \\ 2 & 3 & 7 & -3 & -2 \end{bmatrix}$. Find the following sub-spaces associated with A.

(a) The column space of A is the span of

$$\{(1,\ 0,\ -1/2,\ \underline{\quad}\ ),\ (0,\ 1,\ \underline{\quad},\ \underline{\quad})\}.$$

(b) The row space of A is the span of

$$\{(1,\ 0,\ \underline{\quad},\ 0,\ \underline{\quad}),\ (0,\ 1,\ 1,\ \underline{\quad},\ 0)\}.$$

(c) The nullspace of A is the span of

$$\{(\underline{\quad},\ -1,\ 1,\ 0,\ 0),\ (0,\ 1,\ 0,\ \underline{\quad},\ 0),$$
$$(1,\ 0,\ 0,\ \underline{\quad},\ \underline{\quad})\}.$$

(d) The nullspace of A^t is the span of

$$\{(\underline{\quad},\ \underline{\quad},\ 1,\ 0),\ (-1/2,\ \underline{\quad},\ 0,\ 1)\}.$$

(6) Let $A = \begin{bmatrix} 1 & -2 & -1 & 3 & 2 \\ -2 & 4 & 2 & -6 & -4 \\ 5 & -10 & -1 & 15 & 0 \\ 3 & -6 & 1 & 9 & -4 \\ 3 & -6 & -1 & 9 & 1 \\ 0 & 0 & 2 & 0 & -5 \end{bmatrix}.$

(a) The nullspace of A is the span of

$$\{(2,\ 1,\ 0,\ 0,\ 0),\ (\underline{\quad},\ 0,\ 0,\ 1,\ 0),$$
$$(\underline{\quad},\ 0,\ \underline{\quad},\ 0,\ 1)\}.$$

(b) The row space of A is the span of

$$\{(1,\ -2,\ 0,\ \underline{\quad},\ -1/2),\ (0,\ 0,\ 1,\ \underline{\quad},\ \underline{\quad})\}.$$

(c) The column space of A is the span of

$$\{(1,\ -2,\ 0,\ \underline{\quad},\ 1/2,\ \underline{\quad}),\ (0,\ 0,\ 1,\ \underline{\quad},\ \underline{\quad},\ 1/2)\}.$$

(d) The left nullspace of A is the span of

$$\{(2,\ 0,\ -1,\ 1,\ 0,\ 0),\ (\underline{\quad},\ 1,\ 0,\ 0,\ 0,\ 0),\ (-1/2,\ 0,\underline{\quad},\ 0,\ 1,\ 0),$$
$$(\underline{\quad},\ 0,\ -1/2,\ 0,\ 0,\ 1)\}.$$

(7) Let $A = \begin{bmatrix} 1 & 2 & 1 \\ 2 & 4 & 3 \\ 3 & 6 & 4 \end{bmatrix}.$

(a) Fill in coordinates of the following vector **x** so that it is perpendicular to the rowspace of A. Answer: $\mathbf{x} = (10, \underline{\hspace{0.5cm}}, \underline{\hspace{0.5cm}})$.

(b) Fill in coordinates of the following vector **y** so that it is perpendicular to the columnspace of A. Answer: $\mathbf{y} = (3, \underline{\hspace{0.5cm}}, \underline{\hspace{0.5cm}})$.

(8) In this exercise we prove a slightly different version of the *fundamental theorem of linear algebra* than the one given in Theorem 22.1.4. Here we work with real inner product spaces and the scope of the result is not restricted to finite dimensional spaces, but we must assume that the linear map with which we are dealing has a transpose.

Theorem 22.2.1 (Fundamental theorem of linear algebra). *Suppose that V and W are arbitrary real inner product spaces and that the linear transformation $T\colon V \to W$ has a transpose. Then*

$$\ker T = (\operatorname{ran} T^t)^\perp.$$

We prove the preceding theorem. For each step in the proof give the appropriate reason. Choose from the following list.

DK	Definition of "Kernel"
DO	Definition of "Orthogonal"
DOC	Definition of "Orthogonal Complement"
DR	Definition of "Range"
DT	Definition of "Transpose"
H	Hypothesis
PIP	Elementary Property of Inner Products

Proof. We must show two things: (i) $\ker T \subseteq (\operatorname{ran} T^t)^\perp$ and (ii) $(\operatorname{ran} T^t)^\perp \subseteq \ker T$.

To prove (i) we suppose that $x \in \ker T$ and prove that $x \in (\operatorname{ran} T^t)^\perp$. Let v be a vector in $\operatorname{ran} T^t$. Then there exists a vector w in W such that $v = T^t w$ (reason: \underline{\hspace{1.5cm}}). We compute the inner product of x and v.

$$\langle x, v \rangle = \langle x, T^t w \rangle$$
$$= \langle Tx, w \rangle \qquad \text{(reason: \underline{\hspace{1.5cm}})}$$
$$= \langle 0, w \rangle \qquad \text{(reason: \underline{\hspace{1.5cm}} and \underline{\hspace{1.5cm}})}$$
$$= 0 \qquad \text{(reason: \underline{\hspace{1.5cm}})}$$

From this we infer that $x \perp v$ (reason: \underline{\hspace{1.5cm}}) and consequently that $x \in (\operatorname{ran} T^t)^\perp$ (reason: \underline{\hspace{1.5cm}}).

To prove the converse we suppose that $x \in (\operatorname{ran} T^t)^{\perp}$ and show that $x \in \ker T$. We know that $x \perp \operatorname{ran} T^t$ (reason: _____ and _____). If $w \in W$ then the vector $T^t w$ belongs to $\operatorname{ran} T^t$ (reason: _____); so $x \perp T^t w$ for all $w \in W$. Thus for all $w \in W$

$$0 = \langle x, T^t w \rangle \qquad (\text{reason: } \underline{\hspace{1cm}})$$
$$= \langle Tx, w \rangle \qquad (\text{reason: } \underline{\hspace{1cm}})$$

It follows from this that $Tx = 0$ (reason: _____). That is, $x \in \ker T$ (reason: _____). $\square$

(9) The matrix $\begin{bmatrix} x & y & z \\ y & 1 & x \end{bmatrix}$ has rank one if and only if the point (x, y, z) lies on the parametrized curve $\mathbf{r}(t) = ($ _____, t, _____$)$ in $\mathbb{R}^3$. *Hint.* Use Proposition 22.1.3.

(10) Let A be the 3×4 matrix whose nullspace is the subspace of $\mathbb{R}^4$ spanned by the vectors $(1, 0, 1, 0)$ and $(0, 1, 1, 0)$. Then the vectors $($ _____, _____, _____, $0)$ and $(0,$ _____, _____, _____ $)$ form an orthonormal basis for the row space of A.

(11) Let $T \colon \mathbb{R}^3 \to \mathbb{R}^2$ be the linear transformation whose matrix representation is $\begin{bmatrix} 1 & 0 & 2 \\ 1 & 1 & 4 \end{bmatrix}$ and let $\mathbf{x} = (5, 4, -9)$.

   (a) Find $\mathbf{u} \in \ker T$ and $\mathbf{v} \in \operatorname{ran} T^t$ such that $\mathbf{x} = \mathbf{u} + \mathbf{v}$.
      Answer: $\mathbf{u} = ($ _____, _____, _____ $)$ and $\mathbf{v} = ($ _____, _____, _____ $)$.
   (b) Find $\mathbf{y} \in \operatorname{ran} T$ and $\mathbf{z} \in \ker T^t$ such that $T\mathbf{x} = \mathbf{y} + \mathbf{z}$.
      Answer: $\mathbf{y} = ($ _____, _____ $)$ and $\mathbf{z} = ($ _____, _____ $)$.

22.3 Problems

(1) Let $T \colon \mathbb{R}^5 \to \mathbb{R}^4$ be a linear transformation whose matrix representation is

$$[T] = \begin{bmatrix} 1 & 2 & 0 & -5 & 3 \\ -2 & -4 & 3 & 1 & 0 \\ -1 & -2 & 3 & -4 & 3 \\ 1 & 2 & 3 & -14 & 9 \end{bmatrix}.$$

Gauss-Jordan reduction applied to $[T]$ yields the matrix

$$B = \begin{bmatrix} 1 & 2 & 0 & -5 & 3 \\ 0 & 0 & 1 & -3 & 2 \\ 0 & 0 & 0 & 0 & 0 \\ 0 & 0 & 0 & 0 & 0 \end{bmatrix}$$ and applied to the transpose of $[T]$ yields

$$C = \begin{bmatrix} 1 & 0 & 1 & 3 \\ 0 & 1 & 1 & 1 \\ 0 & 0 & 0 & 0 \\ 0 & 0 & 0 & 0 \\ 0 & 0 & 0 & 0 \end{bmatrix}.$$

(a) From the matrices above we can read off the dimension of the range of T and write down a basis for it. Explain carefully.

(b) From the matrices above we can read off the dimension of the range of the transpose of T and write down a basis for it. Explain carefully.

(c) From the matrices above we can write down two equations that a vector (v, w, x, y, z) must satisfy to be in the kernel of T. Explain carefully. What are the equations? Also explain carefully how we obtain from these equations the dimension of the kernel of T and find a basis for it. Carry out the calculation you describe.

(d) From the matrices above we can write down two equations that a vector (w, x, y, z) must satisfy to be in the kernel of the transpose of T. Explain carefully. What are the equations? Also explain carefully how we obtain from these equations the dimension of the kernel of T^t and find a basis for it. Carry out the calculation you describe.

(2) Let $T \colon \mathbb{R}^6 \to \mathbb{R}^3$ be a linear transformation whose matrix representation is

$$[T] = \begin{bmatrix} 1 & 2 & 0 & 2 & -1 & 1 \\ 3 & 6 & 1 & 1 & -2 & 1 \\ 5 & 10 & 1 & 5 & -4 & 3 \end{bmatrix},$$

Gauss-Jordan reduction applied to $[T]$ yields the matrix $B =$
$$\begin{bmatrix} 1 & 2 & 0 & 2 & -1 & 1 \\ 0 & 0 & 1 & -5 & 1 & -2 \\ 0 & 0 & 0 & 0 & 0 & 0 \end{bmatrix}$$, and applied to the transpose of $[T]$ yields

$$C = \begin{bmatrix} 1 & 0 & 2 \\ 0 & 1 & 1 \\ 0 & 0 & 0 \\ 0 & 0 & 0 \\ 0 & 0 & 0 \\ 0 & 0 & 0 \end{bmatrix}.$$

(a) From the matrices above we can read off the dimension of the range of T and write down a basis for it. Explain carefully.

(b) From the matrices above we can read off the dimension of the range of the transpose of T and write down a basis for it. Explain carefully.

(c) From the matrices above we can write down two equations that a vector (u, v, w, x, y, z) in $\mathbb{R}^6$ must satisfy to be in the kernel of T. Explain carefully. What are the equations? Also explain carefully how we obtain from these equations the dimension of the kernel of T and find a basis for it. Carry out the calculation you describe.

(d) From the matrices above we can write down two equations that a vector (x, y, z) in $\mathbb{R}^3$ must satisfy to be in the kernel of the transpose of T. Explain carefully. What are the equations? Also explain carefully how we obtain from these equations the dimension of the kernel of T^t and find a basis for it. Carry out the calculation you describe.

(3) Let $T: \mathbb{R}^6 \to \mathbb{R}^5$ be the linear transformation whose matrix representation is

$$A = [T] = \begin{bmatrix} 1 & 2 & -1 & -2 & 3 & 0 \\ 0 & 0 & 0 & 1 & -1 & 2 \\ 2 & 4 & -2 & -4 & 7 & -4 \\ 0 & 0 & 0 & -1 & 1 & -2 \\ 3 & 6 & -3 & -6 & 7 & 8 \end{bmatrix}.$$

You may use the following fact: the reduced row echelon forms of the augmented matrix $[A \,|\, \mathbf{b}]$ and of A^t are

$$B = \begin{bmatrix} 1 & 2 & -1 & 0 & 0 & 8 & 3b_1 + 2b_2 - b_3 \\ 0 & 0 & 0 & 1 & 0 & -2 & -2b_1 + b_2 + b_3 \\ 0 & 0 & 0 & 0 & 1 & -4 & -2b_1 + b_3 \\ 0 & 0 & 0 & 0 & 0 & 0 & b_2 + b_4 \\ 0 & 0 & 0 & 0 & 0 & 0 & -7b_1 + 2b_3 + b_5 \end{bmatrix} \quad \text{and}$$

$$C = \begin{bmatrix} 1 & 0 & 0 & 0 & 7 \\ 0 & 1 & 0 & -1 & 0 \\ 0 & 0 & 1 & 0 & -2 \\ 0 & 0 & 0 & 0 & 0 \\ 0 & 0 & 0 & 0 & 0 \\ 0 & 0 & 0 & 0 & 0 \end{bmatrix}, \quad \text{respectively.}$$

Suppose that $\mathbf{x} = \begin{bmatrix} t \\ u \\ v \\ w \\ y \\ z \end{bmatrix}$.

(a) What are the free variables of the system $A\mathbf{x} = \mathbf{0}$ and which are the lead variables? How do you know?

(b) What is the rank of A? Why?

(c) Write a general solution to the homogeneous equation $A\mathbf{x} = \mathbf{0}$ as a linear combination of vectors in $\mathbb{R}^6$ using the free variables as coefficients. Explain.

(d) Explain how to find the dimension of and a basis for the kernel of T. Do so.

(e) Explain how to find the dimension of and a basis for the range of T. Do so.

(f) What conditions must the vector $\mathbf{b} = \begin{bmatrix} b_1 \\ b_2 \\ b_3 \\ b_4 \\ b_5 \end{bmatrix}$ satisfy in order that the nonhomogeneous equation $A\mathbf{x} = \mathbf{b}$ have solutions?

(g) Find, if possible, the general solution to the nonhomogeneous equa-

tion $A\mathbf{x} = \begin{bmatrix} 1 \\ -3 \\ 2 \\ 3 \\ 3 \end{bmatrix}$. (Write your answer as a general solution to the

homogeneous equation plus a particular solution.)

(h) Explain how to find the dimension of and a basis for the range of T^t. Do so.

(i) Explain how to find the dimension of and a basis for the kernel of T^t. Do so.

(4) Prove the three corollaries to the *fundamental theorem of linear algebra* 22.1.4 for complex inner product spaces.

22.4 Answers to Odd-Numbered Exercises

(1) (a) $9 - 3i$, $3 - i$
  (b) 3, -2
  (c) -2, -1
  (d) $-9 - 3i$, $-3 - i$

(3) (a) 2, 1
  (b) $2x + y - z = 0$
  (c) 2
  (d) 8, -22, 2, -8
  (e) 4
  (f) -4, 6

(5) (a) $\dfrac{1}{2}$, $\dfrac{1}{2}$, $\dfrac{3}{2}$
  (b) 2, -1, -1
  (c) -2, 1, 0, 1
  (d) $\dfrac{1}{2}$, $-\dfrac{1}{2}$, $-\dfrac{3}{2}$

(7) (a) -5, 0
  (b) 3, -3

(9) t^2, t^3

(11) (a) 6, 6, -3, -1, -2, -6
  (b) -13, -27, 0, 0

Chapter 23

ORTHOGONAL PROJECTIONS

23.1 Background

Topics: orthogonal and unitary operators, orthogonal projections

Definition 23.1.1. A linear operator $T\colon V \to V$ on a real inner product space is ORTHOGONAL if it is invertible and $T^t = T^{-1}$. A matrix is ORTHOGONAL if it is the representation of an orthogonal operator. An operator T on a complex inner product space V is UNITARY if it is invertible and $T^* = T^{-1}$. A matrix is UNITARY if it is the representation of a unitary operator.

The definitions for *orthogonal* and *unitary* matrices given above differ from the ones offered in 18.1.4. In Problem 8 you are asked to show that in both cases the definitions are equivalent.

Definition 23.1.2. An operator T on an inner product space V is an ISOMETRY if it preserves the distance between vectors. Equivalently, T is an isometry if $\|T\mathbf{x}\| = \|\mathbf{x}\|$ for every $\mathbf{x} \in V$.

Definition 23.1.3. Let V be an inner product space and suppose that it is the vector space direct sum of M and N. Then the projection $E_{MN}\colon V \to V$ is an ORTHOGONAL PROJECTION if $M \perp N$ (that is if V is the orthogonal direct sum of M and N).

Proposition 23.1.4. *A projection E on a complex inner product space V is an orthogonal projection if and only if E is self-adjoint. On a real inner product space a projection is orthogonal if and only if it is symmetric.*

For a proof of (the real case of) this result see exercise 4.

Definition 23.1.5. We say that a linear operator T on an inner product space V is POSITIVE (and write $T \geq 0$) if $\langle Tx, x \rangle \geq 0$ for all $x \in V$. If S and T are two linear operators on V, we say that Q DOMINATES (or MAJORIZES) P if $Q - P \geq 0$. In this case we write $P \leq Q$.

23.2 Exercises

(1) The matrix representation of the orthogonal projection operator taking $\mathbb{R}^3$ onto the plane $x + y + z = 0$ is
$$\begin{bmatrix} & & \\ & & \\ & & \end{bmatrix}.$$

(2) Find a vector $u = (u_1, u_2, u_3)$ in $\mathbb{C}^3$ such that the matrix
$$\begin{bmatrix} \frac{1}{m} & \frac{-1}{m} & u_1 \\ \frac{i}{m} & \frac{-i}{m} & u_2 \\ \frac{1-i}{m} & \frac{1-i}{m} & u_3 \end{bmatrix}$$
is unitary.

Answer: $u = \dfrac{1}{\sqrt{n}} (2+ai, 3-bi, c+di)$ where $a = $ ___, $b = $ ___, $c = $ ___, $d = $ ___, $m = $ ___, and $n = $ _____.

(3) The orthogonal projection of the vector $(2, 0, -1, 3)$ on the plane spanned by $(-1, 1, 0, 1)$ and $(0, 1, 1, 1)$ in $\mathbb{R}^4$ is $\frac{1}{5}(1, a, b, a)$ where $a = $ ___ and $b = $ ___. The matrix that implements this orthogonal projection is
$$\frac{1}{5}\begin{bmatrix} c & -d & e & -d \\ -d & e & d & e \\ e & d & c & d \\ -d & e & d & e \end{bmatrix}$$
where $c = $ ___, $d = $ ___, and $e = $ ___.

(4) Let E be a projection operator on a real inner product space. Below we prove (the real case of) Proposition 23.1.4: that E is an orthogonal projection if and only if $E = E^t$. Fill in the missing reasons and steps. Choose reasons from the following list.

(DK)   Definition of "kernel".

(DL)   Definition of "linear".

(DO)   Definition of "orthogonal".

(DOP) Definition of "orthogonal projection".

(DT)   Definition of "transpose".

(GPa)   Problem 5 in chapter 5.

(GPb)   Problem 1 in chapter 10.

(GPc)   Problem 5 in chapter 17.

(GPd)   Problem 6 in chapter 21.

(H1)   Hypothesis that $M \perp N$.

(H2)   Hypothesis that $E = E^t$.

(PIP)   Elementary property of Inner Products.

(Ti)   Theorem 10.1.2, part (i).

(Tiii)   Theorem 10.1.2, part (iii).

(Tiv)   Theorem 10.1.2, part (iv).

(VA)   Vector space arithmetic (consequences of vector space axioms, etc.)

Let $E = E_{MN}$ be a projection operator on a real inner product space $V = M \oplus N$. Suppose first that E is an orthogonal projection. Then $M \perp N$ (reason: _____). If $\mathbf{x}$ and $\mathbf{y}$ are elements in V, then there exist unique vectors $\mathbf{m}, \mathbf{p} \in M$ and $\mathbf{n}, \mathbf{q} \in N$ such that $\mathbf{x} = \mathbf{m} + \mathbf{n}$ and $\mathbf{y} = \mathbf{p} + \mathbf{q}$ (reason: _____). Then

$$
\begin{aligned}
\langle E\mathbf{x}, \mathbf{y} \rangle &= \langle E(\mathbf{m} + \mathbf{n}), \mathbf{p} + \mathbf{q} \rangle \\
&= \langle E\mathbf{m} + E\mathbf{n}, \mathbf{p} + \mathbf{q} \rangle \quad \text{(reason:_____ and _____)} \\
&= \langle \mathbf{0} + E\mathbf{n}, \mathbf{p} + \mathbf{q} \rangle \quad \text{(reason:_____ and _____)} \\
&= \langle E\mathbf{n}, \mathbf{p} + \mathbf{q} \rangle \quad \text{(reason:_____)} \\
&= \langle \mathbf{n}, \mathbf{p} + \mathbf{q} \rangle \quad \text{(reason:_____ and _____)} \\
&= \langle \mathbf{n}, \mathbf{p} \rangle + \langle \mathbf{n}, \mathbf{q} \rangle \quad \text{(reason:_____)} \\
&= \mathbf{0} + \langle \mathbf{n}, \mathbf{q} \rangle \quad \text{(reason:_____ and _____)} \\
&= \langle \mathbf{m}, \mathbf{q} \rangle + \langle \mathbf{n}, \mathbf{q} \rangle \quad \text{(reason:_____ and _____)} \\
&= \langle \mathbf{x}, \mathbf{q} \rangle \quad \text{(reason:_____)} \\
&= \langle \mathbf{x}, E\mathbf{q} \rangle \quad \text{(reason:_____ and _____)} \\
&= \langle \mathbf{x}, \mathbf{0} + E\mathbf{q} \rangle \quad \text{(reason:_____)} \\
&= \langle \mathbf{x}, E\mathbf{p} + E\mathbf{q} \rangle \quad \text{(reason:_____ and _____)} \\
&= \langle \mathbf{x}, E(\mathbf{p} + \mathbf{q}) \rangle \quad \text{(reason:_____ and _____)} \\
&= \langle \mathbf{x}, E^{tt}\mathbf{y} \rangle \quad \text{(reason:_____)} \\
&= \langle E^t\mathbf{x}, \mathbf{y} \rangle \quad \text{(reason:_____)}.
\end{aligned}
$$

From this we conclude that $E = E^t$ (reason: _____).

Conversely, suppose that $E = E^t$. To show that $M \perp N$ it is enough to show that $\mathbf{m} \perp \mathbf{n}$ for arbitrary elements $\mathbf{m} \in M$ and $\mathbf{n} \in N$.

$$\langle \mathbf{n}, \mathbf{m} \rangle = \langle E\mathbf{n}, \mathbf{m} \rangle \qquad \text{(reason:_____ and _____)}$$
$$= \langle \mathbf{n}, E^t\mathbf{m} \rangle \qquad \text{(reason:_____)}$$
$$= \langle \mathbf{n}, E\mathbf{m} \rangle \qquad \text{(reason:_____)}$$
$$= \langle \mathbf{n}, \mathbf{0} \rangle \qquad \text{(reason:_____ and _____)}$$
$$= 0 \qquad \text{(reason:_____)}$$

Thus $\mathbf{m} \perp \mathbf{n}$ (reason: _____).

Note: Of course, the complex inner product space version of the preceding result says that if E is a projection operator on a complex inner product space, then E is an orthogonal projection if and only if it is self-adjoint.

(5) Let P be the orthogonal projection of $\mathbb{R}^3$ onto the subspace spanned by the vectors $(1, 0, 1)$ and $(1, 1, -1)$. Then $[P] = \dfrac{1}{6} \begin{bmatrix} a & b & c \\ b & b & -b \\ c & -b & a \end{bmatrix}$ where $a = \underline{\quad}$, $b = \underline{\quad}$, and $c = \underline{\quad}$.

(6) Find the image of the vector $\mathbf{b} = (1, 2, 7)$ under the orthogonal projection of $\mathbb{R}^3$ onto the column space of the matrix $A = \begin{bmatrix} 1 & 1 \\ 2 & -1 \\ -2 & 4 \end{bmatrix}$.

Answer: (_____ , _____ , _____).

(7) Let $\mathbf{u} = (3, -1, 1, 4, 2)$ and $\mathbf{v} = (1, 2, -1, 0, 1)$. Then the orthogonal projection of $\mathbf{u}$ onto $\mathbf{v}$ is (_____ , _____ , _____ , _____ , _____).

(8) Let $\mathbf{u} = (8, \sqrt{3}, \sqrt{7}, -1, 1)$ and $\mathbf{v} = (1, -1, 0, 2, \sqrt{3})$. Then the orthogonal projection of $\mathbf{u}$ onto $\mathbf{v}$ is $\dfrac{a}{b}\mathbf{v}$ where $a = \underline{\quad}$ and $b = \underline{\quad}$.

(9) Let $\mathbf{u} = (5, 4, 3, \frac{1}{2})$ and $\mathbf{v} = (1, 2, 0, -2)$. Then the orthogonal projection of $\mathbf{u}$ onto $\mathbf{v}$ is $\dfrac{a}{b}\mathbf{v}$ where $a = \underline{\quad}$ and $b = \underline{\quad}$.

(10) Find the point $\mathbf{q}$ in $\mathbb{R}^3$ on the ray connecting the origin to the point $(2, 4, 8)$ that is closest to the point $(1, 1, 1)$.

Answer: $\mathbf{q} = \dfrac{1}{3}$ (_____, _____, _____).

(11) Let $\mathbf{e}^1 = (\frac{2}{3}, \frac{2}{3}, -\frac{1}{3})$ and $\mathbf{e}^2 = (-\frac{1}{3}, \frac{2}{3}, \frac{2}{3})$ be vectors in $\mathbb{R}^3$. Notice that $\{\mathbf{e}^1, \mathbf{e}^2\}$ is an orthonormal set.

(a) Find a vector $\mathbf{e}^3$ whose first coordinate is positive such that $B = \{\mathbf{e}^1, \mathbf{e}^2, \mathbf{e}^3\}$ is an orthonormal basis for $\mathbb{R}^3$. Answer: $\frac{1}{3}$ (____, ____, ____).

(b) Suppose that $\mathbf{x}$ is a vector in $\mathbb{R}^3$ whose Fourier coefficients with respect to the basis B are: $\langle \mathbf{x}, \mathbf{e}^1 \rangle = -2$; $\langle \mathbf{x}, \mathbf{e}^2 \rangle = -1$; and $\langle \mathbf{x}, \mathbf{e}^3 \rangle = 3$. Then $\mathbf{x} = ($ ____, ____, ____$)$.

(c) Let $\mathbf{y}$ be a vector in $\mathbb{R}^3$ whose Fourier coefficients with respect to B are

$$\langle \mathbf{y}, \mathbf{e}_1 \rangle = \sqrt{8 - \sqrt{37}};$$

$$\langle \mathbf{y}, \mathbf{e}_2 \rangle = \sqrt{5 - \sqrt{13}}; \qquad \text{and}$$

$$\langle \mathbf{y}, \mathbf{e}_3 \rangle = \sqrt{3 + \sqrt{13} + \sqrt{37}}.$$

Then the length of the vector $\mathbf{y}$ is ____.

(d) The orthogonal projection of the vector $\mathbf{b} = (0, 3, 0)$ onto the plane spanned by $\mathbf{e}_1$ and $\mathbf{e}_2$ is $\frac{2}{3}$ (____, ____, ____).

(e) The orthogonal projection of the vector $\mathbf{b} = (0, 3, 0)$ onto the line spanned by $\mathbf{e}_3$ is $\frac{1}{3}$ (____, ____, ____).

(f) What vector do you get when you add the results of the projections you found in parts (d) and (e)? Answer: (____, ____, ____).

23.3   Problems

(1) Prove that an operator $T\colon V \to V$ on a finite dimensional real inner product space V is orthogonal if and only if it is an isometry. Similarly, on a finite dimensional complex inner product space an operator is unitary if and only if it is an isometry.

(2) Prove that an operator $T\colon V \to V$ on a finite dimensional real inner product space V is orthogonal if and only if $T^t T = I$. What is the corresponding necessary and sufficient condition on a finite dimensional complex inner product space for an operator to be unitary?

(3) Show that if an operator U on a complex inner product space is both Hermitian and unitary, then $\sigma(U) \subseteq \{-1, 1\}$.

(4) Let P and Q be orthogonal projections on a real inner product space. Show that their sum $P + Q$ is an orthogonal projection if and only if $PQ = QP = \mathbf{0}$. *Hint.* Use Proposition 23.1.4.

(5) Explain in detail how to find the matrix that represents the orthogonal projection of $\mathbb{R}^3$ onto the plane $x + y - 2z = 0$. Carry out the computation you describe.

(6) Let P and Q be orthogonal projection operators on a real inner product space V.

   (a) Show that the operator PQ is an orthogonal projection if and only if P commutes with Q.
   (b) Show that if P commutes with Q, then

$$\operatorname{ran}(PQ) = \operatorname{ran} P \cap \operatorname{ran} Q.$$

   Hint. To show that $\operatorname{ran} P \cap \operatorname{ran} Q \subseteq \operatorname{ran}(PQ)$ start with a vector $\mathbf{y}$ in $\operatorname{ran} P \cap \operatorname{ran} Q$ and examine $PQ\mathbf{y}$.

(7) Let P and Q be orthogonal projections on an inner product space V. Prove that the following are equivalent:

   (a) $P \leq Q$;
   (b) $\|P\mathbf{x}\| \leq \|Q\mathbf{x}\|$ for all $\mathbf{x} \in V$;
   (c) $\operatorname{ran} P \subseteq \operatorname{ran} Q$;
   (d) $QP = P$; and
   (e) $PQ = P$.

   Hint. First show that (d) and (e) are equivalent. Then show that (a) $\Rightarrow$ (b) $\Rightarrow$ (c) $\Rightarrow$ (d) $\Rightarrow$ (a). To prove that (b) $\Rightarrow$ (c) take an arbitrary element x in the range of P; show that $\|Q\mathbf{x}\| = \|\mathbf{x}\|$ and that consequently $\|(I - Q)\mathbf{x}\| = \mathbf{0}$. To prove that (d) $\Rightarrow$ (a) show that $(I - P)Q$ is an orthogonal projection; then consider $\|(I - P)Q\|^2$.

(8) In 18.1.4 and 23.1.1 the definitions for *unitary* matrices differ. Show that they are, in fact, equivalent. Argue that the same is true for the definitions given for *orthogonal* matrices.

23.4   Answers to Odd-Numbered Exercises

(1) $\dfrac{1}{3} \begin{bmatrix} 2 & -1 & -1 \\ -1 & 2 & -1 \\ -1 & -1 & 2 \end{bmatrix}$

(3) 3, 4, 3, 1, 2

(5) 5, 2, 1

(7) $\dfrac{2}{7}$, $\dfrac{4}{7}$, $-\dfrac{2}{7}$, 0, $\dfrac{2}{7}$

(9) 4, 3

(11) (a) 2, −1, 2
     (b) 1, −3, 2
     (c) 4
     (d) 1, 4, 1
     (e) −2, 1, −2
     (f) 0, 3, 0

Chapter 24

LEAST SQUARES APPROXIMATION

24.1 Background

Topic: least squares approximation.

24.2 Exercises

(1) Let $A = \begin{bmatrix} 1 & 1 \\ 2 & -1 \\ -2 & 4 \end{bmatrix}$.

    (a) Find an orthonormal basis $\{\mathbf{e}^1, \mathbf{e}^2, \mathbf{e}^3\}$ for $\mathbb{R}^3$ such that $\{\mathbf{e}^1, \mathbf{e}^2\}$ spans the column space of A.

$$\mathbf{e}^1 = \tfrac{1}{n}(a,\, b,\, -b)$$
$$\mathbf{e}^2 = \tfrac{1}{n}(b,\, a,\, b)$$
$$\mathbf{e}^3 = \tfrac{1}{n}(b,\, -b,\, -a)$$

    where $a =$ ___ , $b =$ ___ , and $n =$ ___ .

    (b) To which of the four fundamental subspaces of A does $\mathbf{e}^3$ belong?

    Answer: $\mathbf{e}^3$ belongs to the _____ of A.

    (c) What is the least squares solution to $A\mathbf{x} = \mathbf{b}$ when $\mathbf{b} = (1, 2, 7)$?

    Answer: $\widehat{\mathbf{x}} = ($ ___ , ___ $)$.

(2) Find the best least squares fit by a straight line to the following data: $x = 1$ when $t = -1$; $x = 3$ when $t = 0$; $x = 2$ when $t = 1$; and $x = 3$ when $t = 2$.

Answer: $x =$ ____ $+$ ____ t.

(3) At times $t = -2, -1, 0, 1$, and 2 the data $y = 4, 2, -1, 0$, and 0, respectively, are observed. Find the best line to fit this data. Answer: $y = Ct + D$ where $C =$ ____ and $D =$ ____.

(4) The best (least squares) line fit to the data: $y = 2$ at $t = -1$; $y = 0$ at $t = 0$; $y = -3$ at $t = 1$; $y = -5$ at $t = 2$ is $y = -\dfrac{a}{10} - \dfrac{b}{5}t$ where $a =$ ____ and $b =$ ____.

(5) Consider the following data: $y = 20$ when $t = -2$; $y = 6$ when $t = -1$; $y = 2$ when $t = 0$; $y = 8$ when $t = 1$; $y = 24$ when $t = 2$. Find the parabola that best fits the data in the least squares sense.

Answer: $y = C + Dt + Et^2$ where $C =$ ____, $D =$ ____, and $E =$ ____.

(6) Consider the following data: $y = 2$ when $t = -1$; $y = 0$ when $t = 0$; $y = -3$ when $t = 1$; $y = -5$ when $t = 2$. Find the parabola that best fits the data in the least squares sense.

Answer: $y = C + Dt + Et^2$ where $C =$ ____, $D =$ ____, and $E =$ ____.

(7) Find the plane $50z = a + bu + cv$ that is the best least squares fit to the following data: $z = 3$ when $u = 1$ and $v = 1$; $z = 6$ when $u = 0$ and $v = 3$; $z = 5$ when $u = 2$ and $v = 1$; $z = 0$ when $u = 0$ and $v = 0$.

Answer: $a =$ ____; $b =$ ____; $c =$ ____.

(8) Consider the following data: $y = 4$ at $t = -1$; $y = 5$ at $t = 0$; $y = 9$ at $t = 1$.

(a) Then the best (least squares) line that fits the data is $y = c + dt$ where $c =$ ____ and $d =$ ____.

(b) The orthogonal projection of $\mathbf{b} = (4, 5, 9)$ onto the column space of $A = \begin{bmatrix} 1 & -1 \\ 1 & 0 \\ 1 & 1 \end{bmatrix}$ is (____, ____, ____).

(9) The best least squares solution to the following (inconsistent) system

of equations $\begin{cases} u = 1 \\ v = 1 \\ u + v = 0 \end{cases}$ is $u = $ _____ and $v = $ _____.

24.3 Problems

(1) Explain in detail how to use matrix methods to find the best (least squares) solution to the following (inconsistent) system of equations $\begin{cases} u = 1 \\ v = 1 \\ u + v = 0 \end{cases}$. Carry out the computation you describe.

(2) The following data y are observed at times t: $y = 4$ when $t = -2$; $y = 3$ when $t = -1$; $y = 1$ when $t = 0$; and $y = 0$ when $t = 2$.

    (a) Explain how to use matrix methods to find the best (least squares) straight line approximation to the data. Carry out the computation you describe.

    (b) Find the orthogonal projection of $\mathbf{y} = (4, 3, 1, 0)$ on the column space of the matrix

$$A = \begin{bmatrix} 1 & -2 \\ 1 & -1 \\ 1 & 0 \\ 1 & 2 \end{bmatrix}.$$

    (c) Explain carefully what your answer in (b) has to do with part (a).

    (d) At what time does the largest error occur? That is, when does the observed data differ most from the values your line predicts?

24.4 Answers to Odd-Numbered Exercises

(1) (a) 1, 2, 3
    (b) left nullspace
    (c) 1, 2

(3) -1, 1

(5) 2, 1, 5

(7) -6, 73, 101

(9) $\dfrac{1}{3}, \dfrac{1}{3}$

Part 7

SPECTRAL THEORY OF INNER PRODUCT SPACES

Chapter 25

SPECTRAL THEOREM FOR REAL INNER PRODUCT SPACES

25.1 Background

Topics: the spectral theorem for finite dimensional real inner product spaces.

Definition 25.1.1. An operator T on a finite dimensional real inner product space with an orthonormal basis is ORTHOGONALLY DIAGONALIZABLE if there exists an orthogonal matrix that diagonalizes T.

The following theorem (together with its analog for complex spaces) is the fundamental structure theorem for inner product spaces. It says that any symmetric operator on a finite dimensional real inner product space can be written as a linear combination of orthogonal projections. The coefficients are the eigenvalues of the operator and the ranges of the orthogonal projections are the eigenspaces of the operator.

Theorem 25.1.2 (Spectral Theorem for Finite Dimensional Real Inner Product Spaces). *Let T be a symmetric operator on a finite dimensional real inner product space V, and $\lambda_1, \ldots, \lambda_k$ be the (distinct) eigenvalues of T. For each j let M_j be the eigenspace associated with λ_j and E_j be the projection of V onto M_j along $M_1 + \cdots + M_{j-1} + M_{j+1} + \cdots + M_k$.*

Then T is orthogonally diagonalizable, the eigenspaces of T are mutually orthogonal, each E_j is an orthogonal projection, and the following hold:

(i) $T = \lambda_1 E_1 + \cdots + \lambda_k E_k$,
(ii) $I = E_1 + \cdots + E_k$, and
(iii) $E_i E_j = 0$ when $i \neq j$.

25.2   Exercises

(1) Let T be the operator on $\mathbb{R}^3$ whose matrix representation is

$$\begin{bmatrix} \frac{1}{3} & -\frac{2}{3} & -\frac{2}{3} \\ -\frac{2}{3} & \frac{5}{6} & -\frac{7}{6} \\ -\frac{2}{3} & -\frac{7}{6} & \frac{5}{6} \end{bmatrix}.$$

(a) Find the characteristic polynomial and minimal polynomial for T.
 Answer: $c_T(\lambda) = $ _____.
 $m_T(\lambda) = $ _____.
(b) The eigenspace M_1 associated with the smallest eigenvalue λ_1 is the span of $(1, \underline{\quad}, \underline{\quad})$.
(c) The eigenspace M_2 associated with the middle eigenvalue λ_2 is the span of $(\underline{\quad}, \underline{\quad}, -1)$.
(d) The eigenspace M_3 associated with the largest eigenvalue λ_3 is the span of $(\underline{\quad}, 1, \underline{\quad})$.
(e) Find the (matrix representations of the) orthogonal projections E_1, E_2, and E_3 onto the eigenspaces M_1, M_2, and M_3, respectively.

 Answer: $E_1 = \dfrac{1}{m}\begin{bmatrix} a & a & a \\ a & a & a \\ a & a & a \end{bmatrix}$; $E_2 = \dfrac{1}{n}\begin{bmatrix} b & -c & -c \\ -c & a & a \\ -c & a & a \end{bmatrix}$; $E_3 =$

 $\dfrac{1}{2}\begin{bmatrix} d & d & d \\ d & a & -a \\ d & -a & a \end{bmatrix}$ where $a = \underline{\quad}$, $b = \underline{\quad}$, $c = \underline{\quad}$, $d = \underline{\quad}$,
 $m = \underline{\quad}$, and $n = \underline{\quad}$.
(f) Write T as a linear combination of the projections found in (e).
 Answer: $[T] = \underline{\quad} E_1 + \underline{\quad} E_2 + \underline{\quad} E_3$.

(g) Find an orthogonal matrix Q (that is, a matrix such that $Q^t = Q^{-1}$) that diagonalizes T. What is the associated diagonal form Λ of T?

Answer: $Q = \begin{bmatrix} \frac{a}{\sqrt{b}} & \frac{c}{\sqrt{bc}} & 0 \\ \frac{a}{\sqrt{b}} & -\frac{a}{\sqrt{bc}} & \frac{a}{\sqrt{c}} \\ \frac{a}{\sqrt{b}} & -\frac{a}{\sqrt{bc}} & -\frac{a}{\sqrt{c}} \end{bmatrix}$ and $\Lambda = \begin{bmatrix} \lambda & 0 & 0 \\ 0 & \mu & 0 \\ 0 & 0 & \nu \end{bmatrix}$ where

$a =$ ___, $b =$ ___, $c =$ ___, $\lambda =$ ___, $\mu =$ ___, and $\nu =$ ___.

(2) Let T be the operator on $\mathbb{R}^3$ whose matrix representation is

$$\begin{bmatrix} 2 & 2 & 1 \\ 2 & 2 & -1 \\ 1 & -1 & -1 \end{bmatrix}.$$

(a) The eigenspace M_1 associated with the smallest eigenvalue λ_1 is the span of $(1,$ ___, ___$)$.

(b) The eigenspace M_2 associated with the middle eigenvalue λ_2 is the span of $(1,$ ___, ___$)$.

(c) The eigenspace M_3 associated with the largest eigenvalue λ_3 is the span of $(1,$ ___, ___$)$.

(d) Find the (matrix representations of the) orthogonal projections E_1, E_2, and E_3 onto the eigenspaces M_1, M_2, and M_3, respectively.

Answer: $E_1 = \frac{1}{mn}\begin{bmatrix} a & -a & -b \\ -a & a & b \\ -b & b & c \end{bmatrix}$; $E_2 = \frac{1}{m}\begin{bmatrix} a & -a & a \\ -a & a & -a \\ a & -a & a \end{bmatrix}$;

$E_3 = \frac{1}{n}\begin{bmatrix} a & a & d \\ a & a & d \\ d & d & d \end{bmatrix}$ where $a =$ ___, $b =$ ___, $c =$ ___, $d =$ ___,

$m =$ ___, and $n =$ ___.

(e) Write T as a linear combination of the projections found in (d).

Answer: $[T] =$ ___ $E_1 +$ ___ $E_2 +$ ___ E_3.

(f) Find an orthogonal matrix Q (that is, a matrix such that $Q^t = Q^{-1}$) that diagonalizes T. What is the associated diagonal form Λ of T?

Answer: $Q = \begin{bmatrix} \frac{a}{\sqrt{bc}} & \frac{a}{\sqrt{b}} & \frac{a}{\sqrt{c}} \\ -\frac{a}{\sqrt{bc}} & -\frac{a}{\sqrt{b}} & \frac{a}{\sqrt{c}} \\ -\frac{c}{\sqrt{bc}} & \frac{a}{\sqrt{b}} & 0 \end{bmatrix}$ and $\Lambda = \begin{bmatrix} \lambda & 0 & 0 \\ 0 & \mu & 0 \\ 0 & 0 & \nu \end{bmatrix}$ where

$a =$ ___, $b =$ ___, $c =$ ___, $\lambda =$ ___, $\mu =$ ___, and $\nu =$ ___.

25.3   Problem

(1) Let $A = \begin{bmatrix} 1 & -4 & 2 \\ -4 & 1 & -2 \\ 2 & -2 & -2 \end{bmatrix}$.

 (a) Does A satisfy the hypotheses of the *spectral theorem* 25.1.2 for symmetric operators on a finite dimensional real inner product space? Explain.

 (b) Explain how to find an orthogonal matrix that diagonalizes the matrix A. Carry out the computation you describe.

 (c) Explain in careful detail how to write the matrix A in part (b) as a linear combination of orthogonal projections. Carry out the computations you describe.

25.4   Answers to the Odd-Numbered Exercises

(1) (a) $\lambda^3 - 2\lambda^2 - \lambda + 2$ (or $(\lambda + 1)(\lambda - 1)(\lambda - 2)$); $\lambda^3 - 2\lambda^2 - \lambda + 2$ (or $(\lambda + 1)(\lambda - 1)(\lambda - 2)$)

 (b) 1, 1

 (c) 2, −1

 (d) 0, −1

 (e) 1, 4, 2, 0, 3, 6

 (f) −1, 1, 2

 (g) 1, 3, 2, −1, 1, 2

Chapter 26

SPECTRAL THEOREM FOR COMPLEX INNER PRODUCT SPACES

26.1 Background

Topics: the spectral theorem for finite dimensional complex inner product spaces.

Definition 26.1.1. An operator T on a finite dimensional complex inner product space with an orthonormal basis is UNITARILY DIAGONALIZABLE if there exists a unitary matrix that diagonalizes T.

Theorem 26.1.2 (Spectral Theorem for Finite Dimensional Complex Inner Product Spaces). *Let T be a normal operator on a finite dimensional complex inner product space V, and $\lambda_1, \ldots, \lambda_k$ be the (distinct) eigenvalues of T. For each j let M_j be the eigenspace associated with λ_j and E_j be the projection of V onto M_j along $M_1 + \cdots + M_{j-1} + M_{j+1} + \cdots + M_k$. Then T is unitarily diagonalizable, the eigenspaces of T are mutually orthogonal, each E_j is an orthogonal projection, and the following hold:*

(i) $T = \lambda_1 E_1 + \cdots + \lambda_k E_k,$
(ii) $I = E_1 + \cdots + E_k,$ *and*
(iii) $E_i E_j = 0$ *when* $i \neq j$.

Theorem 26.1.3. *Let T be an operator on a finite dimensional complex inner product space V. Then the following are equivalent:*

(1) T *is normal;*

(2) T *is unitarily diagonalizable; and*

(3) V *has an orthonormal basis consisting of eigenvectors of* T.

For a discussion and proofs of the preceding theorems see, for example, [2], page 250*ff.* or [3], page 284*ff.*

26.2   Exercises

(1) Let $A = \begin{bmatrix} 2 & 1+i \\ 1-i & 3 \end{bmatrix}$.

   (a) Use the *spectral theorem* 26.1.2 to write A as a linear combination
   of orthogonal projections.

   Answer: $A = \alpha E_1 + \beta E_2$ where $\alpha = $ _____ , $\beta = $ _____ , $E_1 = \frac{1}{3}\begin{bmatrix} 2 & -1-i \end{bmatrix}$, and $E_2 = \frac{1}{3}\begin{bmatrix} 1 & 1+i \end{bmatrix}$.

   (b) Find a square root of A.

   Answer: $\sqrt{A} = \frac{1}{3}\begin{bmatrix} 4 & 1+i \end{bmatrix}$.

(2) Let T be the operator on $\mathbb{C}^2$ whose matrix representation is $\begin{bmatrix} 0 & 1 \\ -1 & 0 \end{bmatrix}$.

   (a) The eigenspace V_1 associated with the eigenvalue $-i$ is the span of
   $(1, \underline{\hspace{1cm}})$.

   (b) The eigenspace V_2 associated with the eigenvalue i is the span of
   $(1, \underline{\hspace{1cm}})$.

   (c) The (matrix representations of the) orthogonal projections E_1 and
   E_2 onto the eigenspaces V_1 and V_2, respectively, are $E_1 = \begin{bmatrix} a & b \\ -b & a \end{bmatrix}$;

   and $E_2 = \begin{bmatrix} a & -b \\ b & a \end{bmatrix}$ where $a = $ _____ and $b = $ _____.

   (d) Write T as a linear combination of the projections found in (c).

   Answer: $[T] = $ _____ $E_1 + $ _____ E_2.

(e) A unitary matrix U that diagonalizes $[T]$ is $\begin{bmatrix} a & a \\ -b & b \end{bmatrix}$ where $a =$ ____ and $b =$ ____. The associated diagonal form $\Lambda = U^*[T]U$

of $[T]$ is $\begin{bmatrix} & \\ & \end{bmatrix}$.

(3) Let $N = \dfrac{1}{3}\begin{bmatrix} 4+2i & 1-i & 1-i \\ 1-i & 4+2i & 1-i \\ 1-i & 1-i & 4+2i \end{bmatrix}$.

(a) The matrix N is normal because $NN^* = N^*N = \begin{bmatrix} a & b & b \\ b & a & b \\ b & b & a \end{bmatrix}$ where $a =$ ____ and $b =$ ____.

(b) Thus according to the *spectral theorem 26.1.2* N can be written as a linear combination of orthogonal projections. Written in this form $N = \lambda_1 E_1 + \lambda_2 E_2$ where $\lambda_1 =$ ____,

$\lambda_2 =$ ____, $E_1 = \begin{bmatrix} a & a & a \\ a & a & a \\ a & a & a \end{bmatrix}$, and $E_2 = \begin{bmatrix} b & -a & -a \\ -a & b & -a \\ -a & -a & b \end{bmatrix}$

where $a =$ ____ and $b =$ ____.

(c) A unitary matrix U that diagonalizes N is $\begin{bmatrix} a & -b & -c \\ a & b & -c \\ a & d & 2c \end{bmatrix}$ where $a =$ ____, $b =$ ____, $c =$ ____, and $d =$ ____. The asso-

ciated diagonal form $\Lambda = U^*NU$ of N is $\begin{bmatrix} & \\ & \end{bmatrix}$.

(4) Let T be an operator whose matrix representation is $\begin{bmatrix} 1 & 2 \\ -1 & -1 \end{bmatrix}$.

(a) Regarded as an operator on $\mathbb{R}^2$ is T triangulable? ____. As an operator on $\mathbb{R}^2$ is it diagonalizable? ____.

(b) Show that T regarded as an operator on $\mathbb{C}^2$ is diagonalizable by finding numbers c and d such that the matrix $S = \begin{bmatrix} -2 & -2 \\ c & d \end{bmatrix}$ is invertible and $S^{-1}TS$ is diagonal.

Answer: $c =$ ____ and $d =$ ____.

(c) Show that despite being diagonalizable (as an operator on $\mathbb{C}^2$) T is not normal.

$$\text{Answer: } TT^* = \begin{bmatrix} & & \\ & & \\ & & \end{bmatrix} \neq \begin{bmatrix} & & \\ & & \\ & & \end{bmatrix} = T^*T.$$

(d) Explain briefly why the result of part (c) does not contradict Theorem 26.1.3.

---

(5) Let T be the operator on $\mathbb{C}^3$ whose matrix representation is

$$\frac{1}{6} \begin{bmatrix} 8 - i & 5 - 2i & 2 + 4i \\ -5 + 2i & 8 - i & -4 + 2i \\ -2 - 4i & -4 + 2i & 14 + 2i \end{bmatrix}.$$

(a) Find the characteristic polynomial and minimal polynomial for T.

Answer: $c_T(\lambda) = $ _____.

$m_T(\lambda) = $ _____.

(b) The eigenspace M_1 associated with the real eigenvalue λ_1 is the span of $(1, \_\_\_, \_\_\_)$.

(c) The eigenspace M_2 associated with the complex eigenvalue λ_2 with negative imaginary part is the span of $(1, \_\_\_, \_\_\_)$.

(d) The eigenspace M_3 associated with the remaining eigenvalue λ_3 is the span of $(1, \_\_\_, \_\_\_)$.

(e) Find the (matrix representations of the) orthogonal projections E_1, E_2, and E_3 onto the eigenspaces M_1, M_2, and M_3, respectively.

$$\text{Answer: } E_1 = \frac{1}{m} \begin{bmatrix} 1 & -b & bc \\ b & a & -c \\ -bc & -c & d \end{bmatrix}; \quad E_2 = \frac{1}{n} \begin{bmatrix} 1 & b & e \\ -b & a & e \\ e & e & e \end{bmatrix}; \quad E_3 =$$

$$\frac{1}{p} \begin{bmatrix} 1 & -b & -b \\ b & a & a \\ b & a & a \end{bmatrix} \text{ where } a = \_\_\_, b = \_\_\_, c = \_\_\_, d = \_\_\_, e = \_\_\_,$$

$m = \_\_\_, n = \_\_\_,$ and $p = \_\_\_.$

(f) Write T as a linear combination of the projections found in (e).

Answer: $[T] = \_\_\_\_ E_1 + \_\_\_\_ E_2 + \_\_\_\_ E_3.$

(g) Find a unitary matrix U which diagonalizes T. What is the associated diagonal form Λ of T?

Answer: $U = \begin{bmatrix} \frac{a}{\sqrt{bc}} & \frac{a}{\sqrt{b}} & \frac{a}{\sqrt{c}} \\ \frac{d}{\sqrt{bc}} & -\frac{d}{\sqrt{b}} & \frac{d}{\sqrt{c}} \\ -\frac{bd}{\sqrt{bc}} & \frac{e}{\sqrt{b}} & \frac{d}{\sqrt{c}} \end{bmatrix}$ and $\Lambda = \begin{bmatrix} \lambda & 0 & 0 \\ 0 & \mu & 0 \\ 0 & 0 & \nu \end{bmatrix}$ where

$a =$ ___, $b =$ ___, $c =$ ___, $d =$ ___, $e =$ ___, $\lambda =$ ___, $\mu =$ ___, and $\nu =$ ___.

(h) The operator T is normal because $TT^* = T^*T =$
$\frac{1}{6} \begin{bmatrix} a & -bc & 2bc \\ bc & a & -2b \\ -2bc & -2b & d \end{bmatrix}$ where $a =$ ___, $b =$ ___, $c =$ ___, and
$d =$ ___.

(6) Let T be the operator on $\mathbb{C}^3$ whose matrix representation is
$\frac{1}{3} \begin{bmatrix} 5+2i & 2-i & 2-i \\ 2-i & 5-i & 2+2i \\ 2-i & 2+2i & 5-i \end{bmatrix}$.

(a) Find the characteristic polynomial and minimal polynomial for T.
Answer: $c_T(\lambda) =$ _____.
$m_T(\lambda) =$ _____.

(b) The eigenspace M_1 associated with the real eigenvalue λ_1 is the span of $(1,$ ___, ___ $)$.

(c) The eigenspace M_2 associated with the complex eigenvalue λ_2 with negative imaginary part is the span of $($ ___, ___, $-1)$.

(d) The eigenspace M_3 associated with the remaining eigenvalue λ_3 is the span of $($ ___, $-1,$ ___ $)$.

(e) Find the (matrix representations of the) orthogonal projections E_1, E_2, and E_3 onto the eigenspaces M_1, M_2, and M_3, respectively.
Answer: $E_1 = \frac{1}{m} \begin{bmatrix} a & a & a \\ a & a & a \\ a & a & a \end{bmatrix}$; $E_2 = \frac{1}{n} \begin{bmatrix} b & b & b \\ b & c & -c \\ b & -c & c \end{bmatrix}$; $E_3 =$
$\frac{1}{6} \begin{bmatrix} d & -e & -e \\ -e & a & a \\ -e & a & a \end{bmatrix}$ where $a =$ ___, $b =$ ___, $c =$ ___, $d =$ ___,
$e =$ ___, $m =$ ___, and $n =$ ___.

(f) Write T as a linear combination of the projections found in (e).
Answer: $[T] =$ ___ $E_1 +$ ___ $E_2 +$ ___ E_3.

(g) Find an orthogonal matrix Q (that is, a matrix such that $Q^t = Q^{-1}$) that diagonalizes T. What is the associated diagonal form Λ of T?

Answer: $Q = \begin{bmatrix} \frac{a}{\sqrt{b}} & 0 & \frac{c}{\sqrt{bc}} \\ \frac{a}{\sqrt{b}} & \frac{a}{\sqrt{c}} & -\frac{a}{\sqrt{bc}} \\ \frac{a}{\sqrt{b}} & -\frac{a}{\sqrt{c}} & -\frac{a}{\sqrt{bc}} \end{bmatrix}$ and $\Lambda = \begin{bmatrix} \lambda & 0 & 0 \\ 0 & \mu & 0 \\ 0 & 0 & \nu \end{bmatrix}$ where

$a = \underline{\hspace{1em}}$, $b = \underline{\hspace{1em}}$, $c = \underline{\hspace{1em}}$, $\lambda = \underline{\hspace{1em}}$, $\mu = \underline{\hspace{1em}}$, and $\nu = \underline{\hspace{1em}}$.

26.3 Problems

(1) Let N be a normal operator on a finite dimensional complex inner product space V. Show that $\|N\mathbf{x}\| = \|N^*\mathbf{x}\|$ for all $\mathbf{x} \in V$.

(2) Let N be a normal operator on a complex finite dimensional inner product space V. Show that if $\lambda_1, \ldots, \lambda_k$ are the eigenvalues of N, then $\overline{\lambda_1}, \ldots, \overline{\lambda_k}$ are the eigenvalues of N^*.

(3) Let T be as in exercise 4. Show by direct computation that there is no invertible 2×2 matrix $S = \begin{bmatrix} a & b \\ c & d \end{bmatrix}$ of real numbers such that $S^{-1}TS$ is upper triangular.

26.4 Answers to Odd-Numbered Exercises

(1) (a) $1, 4, -1+i, 1, 1-i, 2$
    (b) $1-i, 5$

(3) (a) $\dfrac{8}{3}, \dfrac{2}{3}$

    (b) $2, 1+i, \dfrac{1}{3}, \dfrac{2}{3}$

    (c) $\dfrac{1}{\sqrt{3}}, \dfrac{1}{\sqrt{2}}, \dfrac{1}{\sqrt{6}}, 0, \begin{bmatrix} 2 & 0 & 0 \\ 0 & 1+i & 0 \\ 0 & 0 & 1+i \end{bmatrix}$

(5) (a) $\lambda^3 - 5\lambda^2 + 8\lambda - 6$ (or $(\lambda^2 - 2\lambda + 2)(\lambda - 3)$); $\lambda^3 - 5\lambda^2 + 8\lambda - 6$ (or $(\lambda^2 - 2\lambda + 2)(\lambda - 3)$)
    (b) $i, -2i$
    (c) $-i, 0$

(d) i, i

(e) 1, i, 2, 4, 0, 6, 2, 3

(f) 3, $1 - i$, $1 + i$

(g) 1, 2, 3, i, 0, 3, $1 - i$, $1 + i$

(h) 19, 7, i, 40

BIBLIOGRAPHY

1. Howard Anton and Chris Rorres, *Elementary Linear Algebra: Applications Version*, eighth ed., John Wiley and Sons, New York, 2000. vii, 45, 81
2. William C. Brown, *A second Course in Linear Algebra*, John Wiley, New York, 1988. viii, 175
3. Charles W. Curtis, *Linear Algebra: An Introductory Approach*, Springer, New York, 1984. 175
4. Paul R. Halmos, *Finite-Dimensional Vector Spaces*, D. Van Nostrand, Princeton, 1958. vii, 58
5. Jim Hefferon, *Linear Algebra*, 2006, http://joshua.smcvt.edu/linearalgebra. 45
6. Kenneth Hoffman and Ray Kunze, *Linear Algebra*, second ed., Prentice Hall, Englewood Cliffs, N.J., 1971. vii, 89, 103, 147
7. Nathan Jacobson, *Lectures in Abstract Algebra*, vol. II, Linear Algebra, Springer-verlag, New York, 1984. 49
8. Steven Roman, *Advanced Linear Algebra*, second ed., Springer-Verlag, New York, 2005. viii, 125
9. Gilbert Strang, *Linear Algebra and its Applications*, second ed., Academic Press, New York, 1980. vii, 131

INDEX

Printed in the United States
by Baker & Taylor Publisher Services